ON WIDE SEAS

MARITIME CURRENTS: HISTORY AND ARCHAEOLOGY

ON WIDE SEAS

The US Navy in the Jacksonian Era

CLAUDE BERUBE

THE UNIVERSITY OF ALABAMA PRESS | *Tuscaloosa*

The University of Alabama Press
Tuscaloosa, Alabama 35487-0380
uapress.ua.edu

Typeface: Adobe Caslon

Cover image: Launch of USS *Pennsylvania*, watercolor on paper, 1837; courtesy of Independence Seaport Museum, Philadelphia, Pennsylvania
Cover design: Lori Lynch

Cataloging-in-Publication data is available from the Library of Congress.
ISBN: 978-0-8173-2107-9
E-ISBN: 978-0-8173-9377-9

In memory of my parents,
Gerard and Georgette Berube

I can not refrain from believing that [America] will one day become the first maritime power of the globe. They are born to rule the seas, as the Romans were to conquer the world.

—Alexis de Tocqueville, 1835

Contents

Figures

Preface

Few American presidents cast as large a shadow on the era in which they lived as Andrew Jackson. Even today, in the twenty-first century, Jackson remains a popular topic for historians. Why, then, another book about the 1830s? This book is not about Jackson the president, nor is it about Jackson the individual. Although he remains a popular subject, there are sufficient biographies and studies by those who have spent their entire careers devoted to research and understanding Jackson that this author would simply have been one among many who undertook that enterprise. But with all the books about his presidency, none have addressed the subject in which this author is interested—the role of the Navy during his eight years as the nation's chief executive. What was the Navy's role in the 1830s under Jackson? This work studies the Navy under the prism of the Jackson presidency. At times Jackson is omnipresent—as with the courts-martial or ordering ships on missions. At other times he is absent. And yet at others, such as major operations, he set a clear and concise strategy trusting ship and squadron operations to Navy leadership.

If the author is to be frank, Jackson was never a particularly favorite subject of study as a person or a president. Early on in my research, I mentioned this to a professor in England who reminded me that one of Field Marshall Bernard Montgomery's biographers felt the same way but was told to remember he was a great man. The word *great* is open to interpretation and various definitions. And so it is with Jackson. Jackson personifies an era of American history—good, bad, and ugly. He is a complex individual. While I still have a somewhat jaundiced view of Jackson's behavior, it is his complexity that makes him a perennially compelling subject, worthy of discussion and debate in the ever-changing rearview mirrors of historical study. This work is the first, major rearview of the Navy during Jackson's presidency, but this author neither hopes nor expects that it will be the last. We have, in history, witnessed the dangers of monolithic thinking. Our profession and our collective understanding of history benefit from civil debates. I look forward to those who disagree with the conclusions in this work and know that I will learn from their research and perspectives as I hope my research will inform readers, just as I hope that the legacy of the late Dr. Robert Remini allows me to respectfully find an absence in his work.

I am indebted to the historians who inspired this work. Daniel Walker Howe, whom I have not met, wrote a magnificent book, *What Hath God Wrought: The Transformation of America, 1815–1848*, for which he rightfully won the Pulitzer Prize in 2007. The book was a tour de force of a rapidly changing nation discussing cultural changes, religion, and the Army's activities as it supported the country's push to the Pacific. But, as a naval officer, I was curious why there was no mention of the first naval squadrons to permanently station on distant waters in support of America's growing trade.

Two years after reading Howe's book, I had the opportunity to interview Robert Remini, who had no peer as a Jackson biographer and the great statesmen of that era given the number of works he produced in his fortunately lengthy career. We met for about an hour in his office when he served as historian of the House of Representatives. Dr. Remini was a giant in his field, but his humility, kindness, and decency made this man truly great. I asked him a simple question: "What did Jackson think of the Navy?" His response was concise: "He didn't like it, he didn't understand it, and he didn't use it." He had a clear view when he explained Jackson's relationship with the Navy, but it was a clarity borne of an absence of research on the matter. In his decades of study, he had not gone through the primary sources about the Navy or discussed it. While I could not quantify the first statement, I could explore the last two. And so began my journey into the Navy during Jackson's presidency.

When John Schroeder was the class of 1957 chair of Naval Heritage at the Naval Academy, we discussed his own related work two decades before. John suggested that I pursue the role of Amos Kendall when he was fourth auditor of the Treasury under Jackson. This investigation yielded highly enlightening information, and I am deeply grateful to John for his work, suggestions, and support. Thank you also to Gene Smith and Randy Papadopoulos for their encouragement.

I had previously touched upon the 1830s in my first book about Commodore Charles Stewart as well as an article in Naval Institute *Proceedings* about the ship of the line USS *Pennsylvania* (launched in 1837). Teaching at the US Naval Academy also provided an opportunity to research in a library full of incredibly talented staff, such as Barbara Manvel, and rich in materials. I delved into two treasure troves that had rarely been mined. As a former staffer to two US senators, I remembered the Center for Legislative Archives at the National Archives in Washington, DC. Historical debates on the floor of the House and Senate are available at most large libraries and even online, but those represent only a portion of the policy-making process. The center holds the records of the committees that considered legislation before it reached the floor. To this end, the files from the House and Senate Naval Affairs Committees were extremely valuable.

The second source instrumental to portions of this work were the court-martial records of that era. As I read through sixteen thousand pages of more than three hundred courts-martials, Jackson appeared again and again in approving, denying, or modifying verdicts. I had neither sought nor expected to find this. But the courts-martial revealed a social aspect of the Navy, life aboard the ships at sea or in port, studies of interactions with other navies and cultures, the conditions of the ships, and the character of the men who commanded them or sailed them. It was with this information and data analysis that I could show a different side of the Navy and of Jackson as commander in chief, or rather judge in chief.

With those and other sources, I have with this work attempted to fill a gap in the study of Jackson's presidency. I do so not as a contrarian to the corpus of all the incredible Jackson biographers to whom we are all grateful but to inform the literature of the period, namely of the Navy's role during this period.

I must be (and am) enriched by conversations I have had with the third individual to influence this work—Williamson "Wick" Murray, professor emeritus of Ohio State University, one of this era's most prolific military historians who became a mentor and friend. His guidance, perspectives, and support for this work were instrumental through the nearly decade-long process of answering the question at the heart of this work.

Wick, I learned more about military strategy and history eating pizza and drinking scotch in your basement than any course. You were the one who encouraged—and sometimes demanded—that I return to pursing a doctoral dissertation even at a later stage in my career. You were there every step of the way offering advice, discussing and debating my findings, encouraging me to pursue new paths, sharing a laugh, showing and explaining your military models to me, and telling me not to quit more times that I can count. Thank you.

Finally, I am deeply grateful to the University of Alabama Press, which published this work. Its acquisition editor, Dan Waterman, was a delight to work with. His steady hand throughout the process was greatly appreciated. My editor, Susan Harris, did what all editors do best—make the end product much better than what they received. Thanks to both of you.

Introduction

A Journey toward Maritime Destiny

On December 3, 1828, the Electoral College met to cast their official votes on the presidential election. The incumbent president, John Quincy Adams, earned the support of most northeastern states and Maryland with a total of 83 electoral votes. His opponent, Andrew Jackson, secured the remainder of the states in the West and South with 178 electoral votes. With that vote, Jackson and his supporters achieved the victory denied them in the 1824 election. One historian notes, "When the Democrats captured the government in 1829, they described their victory as the defeat of an elitist, aristocratic rule and the triumph of democracy."[1] It is ironic, therefore, that Jackson had autocratic tendencies and would find himself characterized as a "king" by anti-Jackson newspapers (fig. I.1). A dramatic decrease in the price of a newspaper and an increase in the literacy rate among Americans during this decade combined to give politicians a greater, more immediate national reach. Newspapers grew, in part, as the extension of political candidates and organizations and had a particular viewpoint they expressed either subtly or overtly.

Those who held power in the nation's capital had reason to be concerned about Jackson's disposition and intended actions upon taking office. The tension was palpable throughout Washington in the months leading up to the inauguration. One of the most prominent senators, Daniel Webster, remarked, "Nobody knows what he will do when he does come. . . . My opinion is that when he comes he will bring a breeze with him. Which way it will play, I cannot tell. . . . My fear is stronger than my hope."[2]

On January 19, 1829,[3] Andrew Jackson and his party began their trek from Nashville to Washington, much of it by the nation's waterways.[4] The trip was not his first experience with a steamboat. Fifteen years earlier, during the Battle of New Orleans, he had confiscated a commercial steamboat, the *Enterprise*, to carry supplies between his units.[5]

The decision to travel most of the way to Washington by steamboat was pragmatic. Carriage travel in the wintery backcountry was challenging. Roads were often simply underdeveloped dirt paths that were difficult to traverse even in warm conditions. Traveling by carriage or horseback also would have required finding appropriate lodging—a rarity west of the Appalachian mountain range. By contrast, the steamboat offered Jackson the luxury of space, comfort, and complete protection from the elements. Steamboat travel was also cost effective. In 1815, the cost of transporting one ton of cargo thirty miles to a port city by wagon was nine dollars; for the same nine dollars, one ton of cargo could be transported three thousand miles by ship.[6] Steamboat travel was not without risks, however. Between 1825 and 1830, for example, 273 people died because of exploding steam boilers.[7]

Jackson's presidency witnessed the construction of hundreds of steamboats at riverside dockyards in Cincinnati, Pittsburgh, and Louisville, attesting to the need to support the rapid expansion westward—an expansion

I.1. "King Andrew the First," 1833. A caricature of Andrew Jackson as a despotic monarch. Library of Congress, Prints and Photographs Division.

facilitated by this new platform.[8] It was ironic that the man who was elected largely due to his service as an Army general made most of the trip to the White House by steamboat. In the public's perception, this embrace of naval technology was uncharacteristic for him, but it would serve as a symbol for his relationship with the Navy over the next decade.

Jackson was joined on the journey to Washington by his late wife's nephew Andrew Jackson Donelson; Donelson's wife, Emily (who would serve as Jackson's unofficial first lady); and their son. Donelson had graduated second in the West Point Class of 1820 and had served as Jackson's aide during his term as governor of Florida.[9] Also aboard were Jackson's niece Mary Eastin, a Mrs. Love and her daughter, Henry Lee (brother of later Confederate general Robert E. Lee) and Lee's wife, William Lewis, and Jackson's nephew and adopted son, Andrew Jackson Jr.[10] Lewis had served as Jackson's quartermaster during the War of 1812; was related to John Eaton, with whom Jackson served in the War of 1812, by marriage; and would become the second auditor of the Treasury.[11]

By the time of Jackson's election, steamboats had been in existence and slowly evolving for nearly forty years. John Fitch built the first steamboat in the United States in 1787 and sailed it on the Delaware River during the Constitutional Convention. As with most technologies, the first versions brought with them great hope for the future, but the realities of technological limitations reined in their utility. This pattern would be repeated in Jackson's Navy when some officers and designers attempted to harness steam power onto warships. It was not until Robert Fulton, inspired by his time in London, debuted the *Clermont* in 1807 that the reality of the steamship began to take hold. Though it would be decades before steamboats were efficient and robust enough for transoceanic voyages, the United States was well suited for the shallow-draft platform of steamboats to ply the extensive riverine network west of the Appalachians.

On January 23, Jackson boarded the 133-ton side-wheel steamboat *Pennsylvania* for the remainder of the journey to Pittsburgh, Pennsylvania, with brief stops in Cincinnati, Ohio, and Wheeling, Virginia.[12] As the *Pennsylvania* approached Cincinnati, two other steamers, the *Robert Fulton* and the *Hercules*, joined her. All three boats were lashed together. The ships were adorned with hickory brooms mounted on their bows, making a clear and intentional statement that Jackson was on his way to sweep up the previous administration's filth, harkening back to Dutch admiral Maarten Tromp's attaching a broom to the top of his mast to sweep the English Channel clear of British ships in the seventeenth century.[13]

At Cincinnati, British author and novelist Frances Trollope had the opportunity to witness Jackson's entourage. Steamboats flanked each side of the *Pennsylvania*. Many men stood on the roofs of the two side boats as cannons ashore saluted Jackson's boat. A quarter mile above the town, Trollope reported, the boats came about and made their way to the landing. "When they arrived[,] . . . the side vessels, separating themselves from the center, fell a few feet back, permitting her to approach before them with her honored freight. All this maneuvering was extremely well executed, and really beautiful."[14]

Trollope's perception of a steamboat as a beautiful thing was not universal. Less than a decade later, Secretary of the Navy James Kirke Paulding described the smoke-producing, noisy steamboats as "sea monsters."[15] Around the same time, British artist J. M. W. Turner painted *The Fighting* Temeraire, which portrayed the inelegant and dirty blackness of the steamboat. In this work, the *Temeraire*, one of the ships of the line that fought at Trafalgar, has her sails unfurled, while the light of the sun on the horizon glimmers off her gilding. A small, blackened side-wheel steamboat, whose burning coal darkens the immediate surroundings, tows her up the Thames. The steamboat tows *Temeraire* not to guide her into or out of port but to deliver her to the breakers and the ignominious death that awaited the ships of the Age of Sail.

Though the steamboat era was beginning for the Navy, a conservative sentiment for the traditional ships remained. Gloriously gilded ships such as the *Temeraire* represented the honor, boldness, seamanship, and resiliency of the Romantic era. This loyalty to tradition was one of the reasons that, during Jackson's second administration, the United States would build the USS *Pennsylvania*, the largest ship of the line ever built.[16]

Unlike Paulding and the Romantics, Jackson was a pragmatist who was adept at recognizing trends and identifying their potential impact on the nation. Steam-powered vessels were no exception. His journey to the inauguration represented a future for the country in which its vast riverine system could be exploited to conduct commerce. The power of steamboats would enable farmers and merchants to increase their production, reduce their costs, and expand their markets by decreasing the time and effort needed to transport goods and people.

On January 29, Jackson disembarked at Pittsburgh and continued to Washington by carriage. His steamship voyage totaled 1,128 miles—more than 80 percent of his eventual journey to Washington.[17] He arrived there ready to apply the principles of convenience and efficiency he had experienced on his voyage to his leadership of the Navy.

Reportedly twenty thousand people attended Jackson's inauguration on

March 4.[18] Without the benefit of twentieth-century sound systems, few could have heard Chief Justice John Marshall administer the oath of office to the new president on the east portico of the Capitol. Francis Scott Key, author of "The Star-Spangled Banner," observed, "[This inauguration] is beautiful; it is sublime."[19] The crowd strained to hear President Jackson—likely heard by only a few hundred people—deliver the comparatively short inaugural address in which he spoke about his constitutional duties and acknowledged the limitations of his authority under Article II of the Constitution. He explained that his priority was the elimination of the national debt, a feat never achieved before (nor since). He also proclaimed a desire to "observe toward the Indian tribes . . . a just and liberal policy, and to give human and considerate attention to their rights."[20] Modern historians will note that the ever-contrarian Jackson had deeds that spoke louder than his words in decimating tribes. He devoted 83 words of his 1,100-word speech to his views on the Navy: "The gradual increase of our Navy, whose flag has displayed in distant climes our skill in navigation and our fame in arms; the preservation of our forts, arsenals, and dockyards, and the introduction of progressive improvements in the discipline and science of both branches of our military service are so plainly prescribed by prudence that I should be excused for omitting their mention sooner than for enlarging on their importance."[21] He ended by stating the bulwark of the nation's defense was the national militia, a concept deeply embedded by his experiences. During his presidency, Jackson's views on the Navy and the militia evolved. At the time of his inauguration, however, he recognized that waterborne commerce was the core of America's economy, and he understood that the Navy would protect America's commerce.

The decade epitomized Jackson, or perhaps he epitomized the 1830s, a decade during which democratic populism, a growing adoption of technology, and visions of both maritime destiny and westward expansion dominated many aspects of society but none more so than the Navy.

The Jacksonian era was more than just the man after whom it was named. Certainly, Jackson was the dominant figure of the period, but no single individual could have implemented such pervasive changes in America. During this era, economics, societal trends, and intellectual movements permeated the country. Many of the changes, such as improvements in literacy rates, public education, and medicine, were positive, though the period also saw regressive policies that threatened American Indians and continued to permit slavery in half of the country. Both the period and Jackson himself could be contradictory, advancing broader, noble ideas while simultaneously making significant mistakes in judgment or vision.

In general, however, the Navy of the 1830s did not suffer from the same shortfalls as American society. The Navy was imperfect, but it reflected a burst of enlightenment that would culminate with the creation of the US Naval Academy. The professionalization of its officers set the Navy on the path to innovative and successful operations in the Mexican-American and Civil Wars. It also laid the groundwork for the most significant period of naval thought in the post–Civil War period with the founding of naval organizations and schools, the education reforms of Admiral William B. Luce, and the navalism proposed by such luminaries as Alfred Thayer Mahan and Theodore Roosevelt.

A great deal has been written about Jackson, though the works have focused primarily on his political life, his military exploits, and his wars against Native American tribes. This book is not one of those. For those seeking illumination about his general life and career or the era that he influenced, I recommend works by but not limited to the following: Harry Watson, Daniel Feller, Mark Cheathem, H. W. Brands, Jon Meacham, the late Robert Remini, and others.[22] A new generation is further exploring his presidency and in particular how he treated African Americans and Native Americans. I defer—as should readers—to the expansive network of younger Jacksonian scholars who are focused on diversity and gender issues during his presidency. This book has a different focus.

This work, instead, focuses on a largely untouched aspect of Jackson's administration—that of the Navy. Naval histories touch upon key events during the eight years of his presidency, such as the USS *Potomac* being sent to Sumatra, but many do not provide context for those events. This book provides a survey of the political, intellectual, strategic, operational, and social aspects of the Navy in the 1830s. During Jackson's presidency, these factors merged for the first time in the Navy's brief fifty-year history.[23] This work reveals an emerging naval culture that became self-aware as officers, policy makers, and influential writers alike were drawn into advancing the maritime component of national policies.

1

Naval Inheritance

When Andrew Jackson concluded his two terms as president of the United States in 1837, his farewell address was more than a simple reflection on his tenure and the accomplishments he hoped would define his legacy. His farewell address was also a recognition of maritime circumstances, opportunities for increased activity, and vision that he had not fully recognized when he was first elected. It represented robust, mature insight about the role of the Navy in the country's economic development and defense. It demonstrated a more attentive and mature reflection of the world and the Navy and was more clearly in line with the navalists—those who believed that a peacetime Navy served as a permanent deterrent to aggression and focused on national image, prestige, diplomatic clout, and national greatness.[1] Jackson's farewell address includes a comparatively lengthy treatise on the Navy's purpose and roles. In 1836, with peace reigning and revenue overflowing, he believed that the United States could, year after year, add to the Navy's strength without burdening the people. The Monroe Doctrine of 1823 with its goal of closing the Western Hemisphere to further European colonization was unenforceable given the relatively small size of the American Navy compared to its European counterparts. That task fell to the British Royal Navy, acting on Foreign Secretary George Canning's policies, which had a shared goal to minimize other European involvement in the Western Hemisphere.[2]

The philosophical evolution of Jackson's views occurred in tandem with the emergence of naval self-identity and culture that became the basis of American naval power. In order to become a global maritime power by the end of the nineteenth century, and a superpower in the twentieth century, the United States required more than simply a large fleet. The Navy needed a culture that recognized and embraced modern technology and education. It needed organized specialists who improved the lives of sailors and officers through the standards of physical health as well as some who tended to their spiritual strength.[3] It needed a rising core of officers and like-minded influential civilians who had the capacity for self-reflection, vision, and advocacy.

Changing global and domestic circumstances, as well as an enhanced recognition of the Navy's role in advancing American commerce, intervention, and deterrence, resulted in naval visionaries advocating a nonparochial, internationalist view. This chapter examines the international, domestic, and naval forces that contributed to the state of the Navy by 1828. It also discusses the personal leadership factors, namely the integral and respective career developments of John Quincy Adams and Andrew Jackson, that affected the trajectory of the Navy. An Army general, Jackson had little experience with the Navy outside of the Battle of New Orleans. Years of military action during the War of 1812 and the Seminole War and years of service as a politician and a judge shaped Jackson.[4] Upon his inauguration in March 1829, Jackson inherited a Navy only three decades old. Yet, that Navy was culturally, operationally, and strategically quite different from his more familiar Army. It also lacked experience as, unlike the Army, the Navy had not seen military action in more than a decade, other than engagements against pirates in the West Indies.

Philosophical Influence on the Republic and the Navy

By the start of the 1830s, the United States had established its egalitarian credentials philosophically but not in practice as witnessed with slavery and those who had no voting rights. "No novelty in the United States struck me more vividly during my stay there than the equality of conditions," Alexis de Tocqueville remarked in his magisterial 1835 work *Democracy in America*. Benjamin Franklin and others had proven that American wealth and position could be earned rather than inherited—at least not in the landed aristocratic sense of European states. Certainly, comparatively small merchant houses of the North and the plantations of the South dotted the economic landscape, but equality of conditions meant that practically any free man willing to work or exploit an idea could benefit from a free and capitalistic society. This ability to improve one's social and economic status, to become the equal of those who had inherited land wealth, was a unique feature of the United States, one that contemporaries in Europe could never hope to achieve. The United States, as Tocqueville noted, was, "eminently democratic."[5] Such a democratic mentality imbued its citizens and the Navy with a sense of empowerment. This mentality was also reflected in the young nation's militia, which voted for its military leaders rather than have them appointed. Jackson, elected in the 1790s, was one such leader.

Jacksonian America was, in many ways, the culmination of the democratic republic first conceptualized by Greek philosophers, though slavery and lack of women's rights darkened the ideal.[6] The concepts of individual freedom and the right to challenge authority had evolved from documents such as the Magna Carta through the Mayflower Compact and English Bill of Rights. The US concept of individual freedom and collective governance were derived from Plato, Augustine, Aquinas, Montesquieu, Rousseau, and perhaps most importantly Hobbes and Locke. In 1829, most Americans, excluding the founders, would not know the names of these intellectuals, but they would have been familiar with their ideas, which were intensely debated in the Federalist and Anti-Federalist papers during the battle for ratification of the US Constitution in 1787.[7]

American naval culture reflected national culture, but in some ways, it progressed more quickly and differently. The Navy was a microcosm of a country that had grown more comfortable in its democratic garb. America of the 1830s sought to move beyond the geographical boundaries set by the Atlantic Ocean and the Mississippi River. In expanding westward toward the Rocky Mountains, the United States took on the role of conqueror in some cases and constable in others. The Navy, on the other hand, better reflected the more commercial and enlightened side of the country's identity. But the young democratic government, whose powers resided largely with the states, experienced a change with the more heavy-handed Jackson, who was intent on expanding the power of the national government's executive branch. In response to more authoritarian leadership in the White House, the Navy developed its own unique culture, a culture that gave it an outward-facing, global focus. The Navy believed it had a destiny, and though it took decades to be realized, the concept was born in the Age of Jackson.

The Army represented a constabulary force Jackson would use against the American Indians, with whom he felt no obligation to adhere to treaties.[8] Tocqueville foresaw that "all the great wartime leaders who may arise in the major democratic nations will find it easier to conquer with the aid of their armies than to make their armies live at peace after conquest."[9]

The Navy, by contrast, served to protect commerce when able and to carry out minor punitive actions as needed. Unlike the Army, the Navy had neither the ability nor the intent to carry out expansive engagements during this period. It established no naval outposts other than agents whose purpose was to provide a conduit with the local governing authorities in foreign ports. It neither conquered nor occupied foreign lands, but its ships were involved in attacks on native islanders. Tocqueville devoted five chapters to the nature of armies but only one sentence on the destiny of America's Navy. He certainly would have been familiar with the Navy during his extensive journeys along the coastline. His brevity on the subject suggests that he may have viewed the US Navy as tasked primarily with protecting American commerce rather than as a potentially occupying, aggressive force like the Army.

Beyond philosophical underpinnings, the Army and Navy differed in other ways. As one historian notes, "the specific factors that will inevitably affect the formulation and execution of a grand strategy are easy to describe. Geographical position is perhaps the most obvious and important."[10] The Army found itself constrained by the realities of American geography, while the Navy was unconstrained by the opportunities the oceans presented. With no ability to transport its troops overseas, the US Army of the 1830s remained restricted to the North American continent by four controlling interests: British-controlled Canada to the north; Texas and Mexico to the southwest; Florida (where Native American tribes, primarily Seminoles, remained a dominant force after Spain ceded the region to the United States in 1821) to the south; and the tribal lands that lay beyond Illinois and the Mississippi River to the west.

Between 1830 and 1838, most US Army forts were situated in these unorganized territories, serving as the nation's barrier between its own civilization and a swath of land it viewed as riddled with a godless people. Posts quickly emerged in the Kansas Territory, such as Fort Leavenworth (1827), and in the Oklahoma Territory, such as Fort Gibson and Fort Towson (both 1824) and Camp Washita (1834). These posts served as conduits for the government-enforced Indian expulsion from Florida, Georgia, Tennessee, Alabama, and Mississippi that began in 1830. Army posts, by their nature, encroached upon and held land. "As the military emerged as an agent of American nationalism, it did so in an egalitarian society that rejected privilege and preferred equality of condition and the unfettered pursuit of liberty."[11]

By contrast, during the same period, the US Navy, which was served by long-established yards in Boston, New York, Philadelphia, Washington, and Charleston, held no ground outside the states with the exception of Pensacola in the panhandle of Florida, which did not become a state until 1845. This presence in Pensacola was established to protect the reserves of live oak plantations in the southwest, a key feature of US warships since the designs of Joshua Humphreys in the 1790s.[12] With neither the ability nor the need to hold territory, the Navy's experience was that of near-continuous duty on foreign stations. Foreign ports required and fostered cooperation among naval powers. They also served as havens for merchant ships. Militant action against established ports would have disrupted the flow of commerce, in direct opposition to the purpose for which the Navy existed in the first place.

Such action also threatened the ability of naval ships to operate and remain on distant stations, reliant as they were on local food, water, and goods.

Geography, in terms of times and distance, also favored the Army in terms of communication with the secretary of war or the president. The Army could conduct few military operations or negotiations without direction from Washington because its senior officers were days or at most a few weeks away by horse. US Navy ships operating in the Mediterranean, Pacific, or elsewhere were months from communication with Washington, which provided naval commanders greater latitude and responsibility in representing their country.[13] The complexity of communication with ships and squadrons was, in fact, exponentially greater because they were not always tied to any one port nor were their routes as predictable as those of Army troops marching on a predetermined path from one post to another.

Communication with the Mediterranean was easier than other distant stations since the Navy based its squadron at Port Mahon, Spain, through which all correspondence passed. With several ships in the squadron, at least one ship was always available for dispatches to the United States. In contrast, the Pacific Squadron, comprising only two ships, was forced to rely on passing merchant ships to exchange information or to take the seemingly drastic (or practical) step chosen by its commodore, Charles Stewart, who used his personal funds to finance the construction of his own dispatch boats to keep in communication with Washington.[14]

Deployed naval squadrons were partly immune from the political machinations of Washington as political fortunes rose and fell. The simple factor of distance from policy makers tempered debates about action or inaction. Squadron commodores were free to engage in more fruitful diplomatic relations with their counterparts or heads of state. Left alone to their deliberations, the squadron commodores demonstrated remarkable restraint in the absence of clear and near-continuous direction, though the impetuous actions of some individuals, such as Captain David Porter during the Fajardo Affair (1824) and Captain John Downes in the attack on Quallah Battoo (1832), deviated from those of the more cautious commodores.[15] This freedom with which many Navy commanders operated effectively came to a halt at the end of the Jacksonian era when the introduction of the telegraph in the 1840s made communication much easier.

The Navy's reliance on diplomacy resulted not only from geography but also from simple necessity. Naval ships were more reliant on other nations to conduct operations than was the Army, which was largely self-reliant in North America. Companies of soldiers had wagon trains of supplies for ammunition, clothing, and food. Also, food and water were generally available throughout the areas in which the Army operated during the 1830s. Horses could graze during rest. Shelter in the form of tents was cheap to procure and easy to set up. By contrast, the Navy was encumbered with ships requiring constant attention and maintenance—a leaking hull threatened their survival, and masts and sails were their only means of propulsion and stability.

The crew likewise required support and maintenance. The necessities of life—food and water—were always in finite supply on board a ship. When those stores were depleted, the crew had to rely on the closest port for replenishment. The availability of supplies depended on that port's willingness to do business with the US Navy and, of course, on the Navy's ability to pay for them.

During the Jacksonian era, West Point graduated approximately forty cadets annually, though not all of them went on to a career as an Army officer. Naval officers, by contrast, began serving and training on their first day in the Navy. Nearly all officers entered the service at the rank of midshipman, most of them only in their midteens. The midshipmen received a sparse education, with the chaplain on each ship responsible for conducting classes,[16] while the remainder of their time was focused on learning the fundamentals of ship handling, navigation, and gunnery.

Rare was the naval officer who entered as a lieutenant, although several Jacksonian era captains had entered at this rank because they already possessed significant maritime experience. Commodore Charles Stewart, for example, who received a commission as a lieutenant at the age of nineteen in 1798, had first gone to sea as a cabin boy on a merchant ship at the age of twelve and had risen through the ranks to command his own merchant ship out of Philadelphia before entering the Navy.[17]

Unlike the Army, the Navy did not produce politicians during the nineteenth century. Several presidents from Washington to Jackson served in the military but exclusively in the Army.[18] Presidential family members also served in the Army, such as Martin Van Buren's son Abraham who was aide-de-camp to Major General Macomb during Jackson's presidency. Both Washington and Jackson, for example, had achieved considerable victories at the end of their wars (the American Revolution and War of 1812 respectively). Americans failed to understand or significantly appreciate that the victories at Yorktown and New Orleans would not have been possible without naval support (achieved by the French in the case of the former). While one naval engagement could be romanticized, there were no civilian journalists on board warships to observe, interview, and report on the actual actions.

Army officers had an advantage in politics because their actions were more immediately available to the media. The other, and more likely, reason was that the ranks of "general" and "captain" seemed far different in the public's mind. Generals commanded an Army of the people and militias, whereas captains commanded only a ship. Generals led strategic campaigns and won wars; captains largely engaged single ship actions since, with the exception of the Battle of Lake Erie Squadron, engagements were rare due to the size of the Navy. This disparity was not lost on the Navy, which would rigorously debate the need for an admiral level rank during this decade.

Jackson's familiarity with the Army was therefore understandable. He was a product of that institution and he had helped shape it in the first decades of the American republic. On December 16, 1814, in declaring martial law in New Orleans prior to the final battle of the War of 1812, he had challenged basic constitutional rights.[19] During his presidency, Jackson surrounded himself with former Army officers. John Eaton, for example, served as his first secretary of war. Jackson's secretary and close confidante was his nephew Andrew Jackson Donelson.

Between 1821 and the Civil War (excluding the Mexican-American War), the size of the Army ranged from five thousand to fifteen thousand men, though Jackson authorized the creation of two new regiments—the First Dragoons in 1833 and the Second Dragoons in 1836.[20] His action suggested a need for the professionalization of the armed forces after the abrupt resignation of forty-seven Army officers in 1835 and double that in 1836, as the Second Seminole War began. One of the era's greatest legislators—and political adversary of Jackson—Henry Clay heard from his own son Henry Clay Jr., a cadet in 1830, that "the Army in itself provides no attraction whatever, other than of a certain and independent support."[21] Jackson, with little tolerance for those who eschewed military service, took a more imperial approach than his predecessors. His views carried over into the Van Buren administration as well. To stem the tide of officers leaving their service early, officers were held accountable and sent back to their units to complete their obligation. This was especially enforced with West Point graduates.[22]

Although the Army had been a significant part of his life and captivated his attention—especially with the Second Seminole War—he was not ignorant of what the Navy offered. Although one Jacksonian historian argued that Jackson had little understanding or use for the Navy, evidence suggests otherwise.[23] Jackson was a product of the Army and a promoter of presidential power. The Navy was less familiar to him, but he understood that it had the capability for divergent missions.

The Navy of 1829

The Navy Jackson inherited in March 1829 faced multiple challenges.[24] Arguably, the Navy lacked a central, justifiable mission that would engender comprehensive support from both the public and Congress. This lack of support manifested itself in numerous ways. In 1829, the Navy found itself still in the early stages of global presence on distant stations. The duties and logistics complications associated with establishing new stations were compounded by a rudimentary industrial base, struggles for resources, disparate education, lack of available manpower, and stagnation in innovation.

Prior appropriations supported construction of a new naval hospital (in addition to that at New Orleans), but there was as yet no system in place for the management of sick and disabled seamen. The Navy had also failed to respond to 1828 legislation providing for a better organization of the medical department. The law required a change in how assistant surgeons were admitted to the service and time in grade before being promoted to surgeon. Secretary of the Navy Samuel Southard reported to Congress that, despite its required measures, none had been put into place.[25] Most of Southard's report consisted of concerns with pension funds from both former active-duty naval personnel and privateer seamen. In most navies—even in the twenty-first century—the most significant cost is not platforms and weapons but personnel, including their training, food, medical care, and other benefits such as retirement.

As it had for most of the 1820s, discussion continued in Congress and in the Navy on the merits of an exploring expedition to the Pacific and South Sea to survey coasts, islands, and reefs. In subsequent reports to Congress, but prior to Jackson's inauguration in March, the secretary of the Navy provided amplifying information on the costs associated with an expedition.[26] Southard and others recognized the demonstrable deficiency of American nautical charts. With an estimated two hundred or more islands, and reefs and shoals that did not appear on any chart, such an expedition was necessary to protect the lives and seafaring property of American whalers in the region. Nevertheless, it would be nearly another decade before the multiship expedition would get underway.

The character of officers and sailors suggested that naval discipline had been generally commendable in 1828, but infractions nonetheless required responses. Some suggested a criminal code, a naval school, and a revised body of Navy rules and regulations.[27] All three potential remedies had been suggested to Congress via separate reports.

Conspicuously absent from Southard's 1828 report to Congress was significant mention of global naval operations, despite squadrons operating in the West Indies, Pacific, Mediterranean, and elsewhere.[28] Southard had only one sentence on piracy operations by the Mediterranean Squadron. Otherwise, an appendix noted only a list of vessels and their locations. Two-thirds of the Navy (twenty-three ships) were deployed that year, including one 74-gun ship of the line.

Southard estimated that the Navy's 1829 budget would be more than $3.6 million.[29] The total budget for the US government that year was $17 million. For a young republic with few resources, devoting 20 percent of its budget was out of proportion to its limited ability. Army and Navy expenditures comprised more than half of the federal budget. Interest on the national debt was another 15 percent. Barely 30 percent remained to run the rest of the government, including the postal service, which was the largest organization outside of the military. More than one-third of the Navy's budget was for the pay and subsistence of its personnel. Less than $800,000 was estimated for provisions and repairs of vessels, and another $400,000 was for repairs and improvements of Navy yards. The budget allocated no money for new construction.

The Navy Jackson inherited in 1829 was also a reflection, in part, of the man who had preceded him as president. John Quincy Adams (fig. 1.1) was

1.1. John Quincy Adams by Asher B. Durand. A key figure in the young nation's diplomatic efforts, John Quincy Adams understood the role of the emergent navy as well as anyone. Courtesy of the US Naval Academy Museum.

not only Jackson's predecessor. To Jackson and his followers, Adams was the man who had usurped the presidency itself during one of the most contentions elections in American history. The rivalry extended to how the Navy was managed.

Adams, unlike Jackson, was one of the early American elites.[30] The son of the president who had created the Department of the Navy in 1798, Adams benefited from his father's political prominence. From an early age, Adams served as his father's secretary during assignments in Europe. Adams became the minister to Holland in Washington's second administration at the youthful age of twenty-seven. He served as minister to Russia from 1809 to 1814 and the next three years as the minister to the Court of St. James. When he returned to the United States in 1817, he was nominated for and confirmed as secretary of state under President James Monroe. By the time he ran for president in 1824, he was one of the most well-traveled Americans. Jackson, by contrast, never visited a foreign country. That such a dichotomy of international experiences between the two successive presidents existed underscores the inherent necessity of a navy and its usefulness. Adams recognized this from the outset of his administration, but Jackson also learned this lesson quickly.

Adams had been at his father's side during the undeclared Quasi-War with France, primarily fought in naval engagements throughout the Caribbean. But it was a later engagement that had more impact. "The really important period of my life," he wrote in his diary on May 31, 1820, "began with the British attack upon our Chesapeake frigate, in the summer of 1807."[31] The USS *Chesapeake*, one of America's first six frigates built in the 1790s, had gotten underway under Captain James Barron. Only a few hours out of Norfolk, Virginia, her decks packed with crates of supplies, she was completely unprepared for any engagement. HMS *Leopard* awaited her with the understanding that several British naval deserters were aboard. *Leopard* emptied a broadside into *Chesapeake*. Barron ordered one cannon fired so that he would not have to say that he had surrendered the ship without firing a shot. The *Leopard*'s captain impressed several seamen but rejected Barron's offer to surrender the ship, which might have escalated into an act of war.

The United States eventually declared war five years later, while Adams served as minister to Russia. Adams found a sympathetic ear for the cause of neutral trade, even though Russia, through 1812 at least, was nominally allied with France. Sensing this, Adams managed to convince Czar Alexander to open Russian ports to American goods. This "diplomatic coup," one historian argues, "illuminates the strategic importance Adams attached to ports as hubs of commercial activity."[32] By the time Adams became secretary of state, his views of the importance of the Navy and the nation's merchant fleet in growing the nation had matured, articulated in his first strategic document.

As secretary of state, Adams recognized the limitations of the nation's small fleet compared to the global dominance of the Royal Navy. Still, he knew the nation could either learn from, in some way influence, or exert commercial power regionally. Despite the British Empire's unchallenged role, there were considerable conflicts that involved navies. In the Mediterranean, for example, war raged between the Greeks and Ottoman Empire. A recent study of Adams's foreign policy suggests that, "inspired by the democracies of ancient Greece, thrilled by the republican rhetoric of the Greek rebels, and horrified by Turkish authorizes, Americans [like Adams] took the Greek cause to heart."[33] Despite this, within a decade the Navy was helping to rebuild the Ottoman fleet under Jackson and Van Buren.[34] Elsewhere, revolutions in South America under Simon Bolivar, Jose de San Martin, and Bernardo O'Higgins challenged the faltering Spanish Empire. Understanding the consequences of strategic overreach, Adams believed the young Navy could have a significant impact closer to home. It was, he and Monroe agreed, the destiny of the country to grow.

Adams was already familiar with America's challenges to the south. Shortly after the War of 1812, Adams was the primary negotiator with Spain on establishing a secure border with the empire's Florida territory. With Jackson's victory over the British at the Battle of New Orleans still fresh in the minds of Americans, he was sent to the Georgia-Florida border in 1817. The radically anti–American Indian Jackson sought to settle the issue of security once and for all thus preventing any potential Indian incursions into Georgia. While Adams played diplomat, Jackson executed two British merchants—Robert Armbrister and Alexander Arbuthnot—for allegedly collaborating with Indians.[35] Severely criticized by Secretary of War John C. Calhoun (later to serve as Jackson's vice president during his first term), Adams defended Jackson's actions for four reasons. Jackson's Army neutralized the Indian threat along the Florida border. Second, it stabilized US borders. Third, "it secured the US against a weakening Spain . . . [and] swiftly and decisively project[ed] American force against Seminole, Spanish, and British threats."[36] Such action risked much. Adams, already a seasoned diplomat would have known the risks, but he also saw opportunity for the country. In a letter to Secretary of War John C. Calhoun, Jackson wrote of the incident: "I hope the execution of these Two [*sic*] unprincipled villains will prove an awfull [*sic*] example to the world and convince [Britain] . . . that certain, if slow retribution awaits those unchristian wretches who by false promises delude & excite a Indian tribe to all the horrid deed of savage war."[37] The Spanish and British governments both expressed their concerns to the US secretary of state.

Monroe was less forgiving of Jackson, informing the general that he had exceeded his orders and risked war. Privately, he admonished Jackson. Publicly, however, Monroe had to tread lightly with a man who was likely more popular than the president himself. On January 12, 1819, the House Committee on Military Affairs condemned the executions and debated a formal censure. House Speaker Clay supported censure, but in the end, Jackson's popularity and rising nationalistic fervor won out with 107 members of Congress voting against censure and 100 voting in favor of it.[38]

For Adams, the only way for the Navy to grow was to increase operations closer to home. In the near term, that meant the Florida territory. In the longer term, it meant the Caribbean and the Spanish territories in Central and South America. The policy he articulated would be attributed to the president under whom he served. Announced in 1823, the Monroe Doctrine declared that the Western Hemisphere was no longer open to European colonization and that any interference would be met with American intervention. While bold and audacious, this policy had no practical method of being implemented. The US Navy had insufficient force to conduct a blockade or engage in any squadron, much less fleet, action with the Royal Navy. In addition, it would not have the capability to project forces over water until the invasion of Mexico in 1847 when thirteen thousand American troops landed at Veracruz. It is also possible that, by underwriting the Monroe Doctrine, the Royal Navy blunted the US Navy, reducing its need to grow to defend itself and thus slowing subsequent development until the 1880s.

As president, Adams became an advocate for the Navy, proposing increases in the size of the fleet as well as the establishment of a naval academy.[39] But most of his efforts came to naught. The same populist tide that had overwhelmingly voted for Jackson swept in a new wave of congressmen who referred to themselves not as Democratic-Republicans but formally as Jacksonians. With four years of a determined opposition, Adams managed to increase the size of the Navy but was unable to accomplish much else given a hostile Congress.[40] The same was true after the 1828 presidential election when memories of the corrupt bargain had festered enough to grant Jackson his outright victory and claim the prize denied to him four years earlier.

At just a few days shy of his sixty-eighth birthday, Jackson (fig. 1.2) was inaugurated as the oldest president to first assume the office. His administration inherited international issues, a nation formed by differing political

1.2. President Andrew Jackson (1829–37) by Asher Durand. Courtesy of the US Naval Academy Museum.

philosophies, and a Navy shaped by several wars and only having recently formed permanent squadrons around the globe. But if Jackson inherited issues that formed the country, then the country likewise inherited a president formed by his experiences. Both fueled each other for the next eight years; nowhere was this more evident than with the US Navy.

As Jackson took office, the country offered his administration a realized sense of democratic style of government. The American culture was on the precipice of new ways of education as well as a moral imperative from a new religious revival movement. The American economy was only beginning its surge forward in maritime commerce that would spur wealth and development. And there was Jackson, arguably one of the most popular Americans since Washington, who was ready to assume leadership over it all. His sense (albeit often misplaced) of justice, of populist democracy, and of governmental accountability also began a new era for the Navy.

The Navy, battle tested by three wars, was a beneficiary of this era of activism and democratization. The mood of the country and administration encouraged and enabled the Navy to advance beyond a simple battle force of wooden hulls and iron cannons. The 1830s offered the Navy opportunities for global presence and influence. Naval officers embarked on missions that fostered a new way of thinking. The Navy was ready for a change. It was ready for an era of intellectual awakening.

2

The Intellectual Awakening of the Navy—A Republic of Ideas

Two factors dramatically changed the ability of officers to disseminate their ideas publicly by the 1830s, both based on the wave of American democratization: cost and education. In the 1830s, the price of newspapers declined from six cents to a penny; books that sold in 1830 for two dollars were only twenty-five cents a decade later.[1] As a result of the dramatic price decrease, newspapers, books, and journals became more affordable to the general population and the number of publications rose during Jackson's presidency. The second factor was the impact of the common schools on literacy rates.[2]

The medium of print and the ability of more Americans to read meant that writers—and specifically naval officers—could reach a wide audience for the first time. Lieutenant Charles Wilkes requested the purchase of a lithographic press for the Navy in 1833 for each Navy yard. He estimated that the Navy would save money from rollers, ink, and printer labor. Moreover, since the Navy had to buy French and British charts, it could make its own and correct them immediately.

American scientific journals made their advent in the postwar period in the 1830s and included articles related to naval issues. The *American Journal of Science*, founded in 1817 by Yale University chemistry professor Benjamin Silliman, included articles on the safety of steamboats and studies on the various bodies of water the Navy was exploring. The increase of Navy-related articles in the 1830s was primarily due to the growing use of steamboats and their application for the Navy. Other articles discussed potential explorations.

The increased interest in the Navy may also be a result of Silliman, who on at least one occasion visited the Naval Lyceum in 1834 with Lieutenant James Glynn.[3] Four years later, Glynn served with the South Seas Exploring Expedition.

These and the literary and scientific journals like them were an integral component of advancing the Navy's learning and, in turn, of the broader academic, literary, and generally American audiences in the 1830s. They shared a common goal with the Navy's junior officers—to share ideas and information in a new era of advancing thought and exploration.

The new generation of officers in the 1830s led an intellectual reformation through concepts articulated in publications and organizations. They challenged conventional thinking in the Navy and confronted their own superiors in print, albeit through the relative anonymity of pseudonyms. Their ideas set the Navy on its eventual path of professionalism. They became the ship captains and commodores who later commanded the ships and squadrons of the Mexican-American War and Civil War.

One British historian writes that the Enlightenment's "direct impact on the navies of the day, as opposed to its indirect, long-term influence on the societies which sustained them, seems in truth to have been rather limited."[4] While that may have been true of European navies, it was not applicable to the US Navy. During Jackson's presidency, the Navy experienced its own age of enlightenment.[5] Intellectually curious junior officers, along with like-minded and supportive literary figures, were learning about the world, a world that was only beginning to realize the nation's potential as a future power. The young republic's naval operations stimulated their curiosity.

The Navy, from its establishment in 1798 through 1815, experienced three wars, and its officers and secretaries of the Navy could do little more than react to short-term threats and objectives. The postwar environment at first brought breathing room for the Navy, and, as it built up its fleet beginning in 1816, it finally had the opportunity to think about the world and its emerging role. The environment created by the post–War of 1812 pause in war fighting enabled the junior officers sufficient time, resources, and opportunities to develop a robust correspondence network about naval policies and innovative ideas. Naval officers shared their experiences with one another and collectively stimulated their knowledge and views. They articulated their thoughts through new publications. In reviewing Lieutenant Matthew Maury's *Treatise on Navigation*, editor Edgar Allan Poe observed:

> The spirit of literary improvement has been awakened among the officers of our gallant navy. . . . Hitherto how little have they improved the golden opportunity of knowledge which their distant voyages held forth, and how little have they enjoyed the rich banquet which nature spreads for them in every clime they visit! But the time is coming when, imbued with a taste for science and spirit of research, they will become ardent explorers of the regions in which they sojourn.[6]

This was the first period of robust published naval thought in the United States. New technologies emerged, and new lands were explored. The Navy debated and developed new, better-structured schools. Conceptual changes to the Navy would not have been possible without appropriate and effective dissemination within and beyond the Navy. During the Age of Jackson, this dissemination included unofficial correspondence between officers, articles in new professional journals, like-minded popular literary figures, education to sow the seeds of learning, and organizations whose role was to share ideas among a commonly interested population. This was the Navy's version of the French Enlightenment philosopher Peter Bayle's (1647–1706) concept of a Republic of Letters in which intellectuals corresponded with each other from great distances and shared ideas and discussed them in academies and salons.[7]

Prior to the Age of Jackson, sharing news and information was difficult at best. From the establishment of the Department of the Navy in 1798 through the end of the War of 1812, the Navy's size was inconsistent, its role was ad hoc, and for most of the period, it suffered from the instability caused by perpetual war, either with France, the Barbary states, or England. Beginning in the postwar period, ships (like the new ships of the line) on longer overseas deployments meant that the Navy's midshipmen would become more familiar with one another as a group rather than isolated in smaller ships dispersed throughout the Eastern Seaboard and deployed. In such close quarters, long-term deployments offered the next generation of midshipmen aboard frigates and ships of the line the opportunity to form professional and personal relationships.

These relationships also included the growth of American-born literary figures whose uniquely American work, such James Fenimore Cooper's *Last of the Mohicans*, appeared in the 1820s and 1830s. This new cadre of authors possessed close ties to the nascent Navy. Some, like Cooper, had served as midshipmen prior to the War of 1812, while others had brothers in the service. Authors like Washington Irving found themselves exposed to the Navy when they wrote about its adventures in the War of 1812 in journals and newspapers.[8] In the 1830s, these same authors wrote fictional tales of the sea as well as nonfiction works. Their popular works reached a hungry American audience beyond what the officers themselves could accomplish.

The combination of internally generated ideas among naval officers, their dissemination in professional organizations, and the popular interest they generated led to exploration, such as the South Seas Exploring Expedition, and education with the establishment in 1845 of a naval school. Better educated than the preceding generation of naval officers, these lieutenants and midshipmen immersed themselves in the culture, history, and geography of foreign ports. They witnessed the instability of other governments. As they grew, their ideas took seed and formed the basis of a new outlook for themselves as professional officers, for the Navy, and for the country as a whole.

The first exposure of many midshipmen to the world was through books. The first ship's library was on the ship of the line *Franklin*, which left the United States in 1817 for a three-year Mediterranean deployment. Its library included volumes purchased by Commodore Charles Stewart and a charitable mariner's society. In 1828, Secretary of the Navy Southard issued a circular for all ships to have a library with thirty-seven specific books, including those on trigonometry; algebra; gunnery; histories of Greece, Rome, and England; the laws of nations; and maritime laws; among others. Correspondence, journals, and diaries in later years often provided the officers an opportunity to reflect on each of their ships, the men who commanded them, and the men who served aboard. Writing to a fellow officer in 1835, and reflecting on his service on a ship just two years prior, Lieutenant Samuel Francis Du Pont (fig. 2.1) noted, "I have always looked upon my cruise in the [USS] *Ontario* as the brightest period of my professional life—it was performed after the . . . thoughtlessness of youth had fortunately yielded a little to the stability of

2.1. Samuel Du Pont represented a generation of naval officers trained in Jacksonian Navy and would rise to command the squadrons of the Civil War. Portrait by Daniel Huntington, ca. 1867–68. National Portrait Gallery, Smithsonian Institution; transfer from the Smithsonian American Art Museum; bequest of May DuPont Saulsbury, 1927.

manhood."[9] By the time of his service aboard *Ontario*, Du Pont had been in the Navy for sixteen years with several overseas deployments.

The topics of the officers' correspondence varied during the Age of Jackson, with the location of their friends of primary concern. But the correspondence was not simply idle chatter. Many thoughts or questions, particularly as they involved what other officers might encounter, expanded their awareness of the Navy and the world around them. Echoing Mackay's observation, Lieutenant Garrett J. Pendergrast was concerned about the disease's impact on the other officers of the squadron operating in the sometimes unforgiving climate. Pendergrast thought that Du Pont had been "numbered with these gallant fellows who have sacrificed themselves for their ungrateful country's good."[10] Relationships were integral to forming later correspondence. Later, Rear Admiral Benjamin F. Sands recalled meeting Midshipman Joshua Sands on their first ship in 1828, "We used to while away many hours endeavoring to trace out a relationship."[11]

News of courts-martial also reached across the fleet, albeit slowly. In May 1829, Lieutenant Daniel Mackay awaited news of one trial involving Lieutenant Irving Shubrick.[12] Shubrick had brought Marine lieutenant Joseph Hale of the *Java* on charges of quarrelling and scandalous conduct including "a challenge to personal combat."[13] Found guilty, Hale received a sentence of suspension for twelve months.[14] That the trial occurred in the Mediterranean meant that it could be weeks to months before the other officers learned of the results. In Mackay's case, it is unlikely he learned of it while in the Gulf of Mexico before the loss of the *Hornet*, but that did not prevent him from rendering a judgment. "I take the Marine will be taught a lesson. The officers of that Corps seemed fated to bring about their own destruction and that of the corps. They are idle and of course doing mischief."[15]

Deployments around the globe allowed officers the opportunity to compare notes about the differences of each port, country, and region and the challenges each offered. This, of course, also included interactions with characters. Du Pont, for example, met with Edward John Trelawny at Napoli de Romania—then the capital of the Greeks at the head of the Gulf of Argos—in September 1825; Trelawny had, the year prior, accompanied Lord Byron to Greece during its war for independence. Surgeon's Mate John Wiley in 1831 wrote of the "novel and beautiful Appearance" of Constantinople and admired everything from the minarets to the white buildings of Lesopho Point in a grove of cypresses. "If I had but one city to see, that city should be Constantinople. . . . European cities always have something in common. Here everything is distinct and peculiar."[16] For Chaplain Thomas Lambert, awe of regions manifested itself in the frigate *United States*'s visit to the Holy Land in 1836. From the hill where St. Paul had preached, Lambert "succeeded in breaking off a piece of the rock which I took away with me."[17] But like Wiley, who was enamored of Constantinople, Lambert found himself drawn to Rio de Janeiro, which he believed was "unrivalled in beauty and grandeur."[18]

Descriptions of these regions were not universally positive, especially about the West Indies Station and Gulf of Mexico. Lieutenant Mackay dreaded his assignment to the region immediately upon receiving orders to USS *Hornet*. "I must begin to pack up," he wrote, "and make my little arrangements, not I confess without some foreboding, not exactly of death and yellow fever, but it's a climate I do not like."[19] Lieutenant Pendergrast was also aware of the region's hazards having "read with much regret the accounts of the fatal consequences of yellow fever in the West India Squadron."[20] Other officers observed the man-made dangers, such as the atrocities committed by Caribbean pirates. Others recognized the instability of Mexico and that the Mexican Army was "exceedingly anxious for a melee."[21] Junior officers were also curious about naval policies and broader American politics first engaging with one another and then through periodicals. A few months before his untimely death aboard *Hornet*, Mackay announced that he planned to write an article on naval policy using reports, documents, and state papers and asked his friends still in the United States to send anything of interest in papers and magazines to him for reference.[22]

Some officers directly solicited members of Congress or those close to them. Thomas Lambert, for example, wrote at the request of the officers of the frigate *Brandywine* about a bill on equalizing the pay of naval officers.[23] Members clearly listened to their concerns as evidenced by the chairman of the Committee on Naval Affairs in 1835 informing a junior officer on the status of a bill.[24] Du Pont and his peers found themselves especially attuned to the administration, commenting that "Jackson has done everything in his power to assist us."[25] They were even aware of the appointment of officials like Amos Kendall, who was not part of the Navy but rather served as fourth auditor of the Treasury Department, which held key accounting responsibilities for the Navy.

In the 1830s junior officers sometimes found issues so compelling that they were willing to risk retribution from their commanding officer or commander in chief to influence public policies. Senior officers attempted to stem this either through court-martial proceedings or direct orders. This was the case when Commodore George W. Rodgers, while commanding the Brazil Squadron in 1831–32, issued an order to all his officers that they were "positively forbidden communication either directly or thro' [*sic*] their friends to

2.2. Commodore Charles Stewart. Commissioned in 1798, Stewart was part of a generation that led the 1830s navy. A captain since 1807, Stewart's generation also represented a hindrance to advancement for younger officers. Courtesy of the US Naval Academy Museum.

the Public Prints in the United States or elsewhere, the movements or transactions that may occur in the Squadron."[26]

Many naval officers married into political families, such as with the daughters of congressmen. In Boston, New York, Philadelphia, and other port cities with established American blue-blood families, naval officers could identify potential spouses. Stewart (fig. 2.2), for example, met and married his wife Delia Tudor—the "Belle of Boston"—a few weeks prior to deploying with *Constellation* in 1812. Of the following decades, Steedman recounted that "Norfolk was the Garden of Eden for naval officers," given that so many had married there.[27]

The diversity and depth of topics in the correspondence of the naval officers during the 1830s suggest a growing naval intellectual community. Fortunately, this awareness did not remain isolated to private correspondence between the officers but evolved as very public debates in the new medium of naval and other journals. Publications became more widespread in the 1830s and, with it, greater opportunities to share ideas. Naval and other publications were integral to the reform movement among officers. Until this time, articles about the Navy in journals—particularly by serving officers—were sporadic and were usually simple accounts of naval actions and hagiographic biographical essays of early naval heroes. New military-specific publications emerged such as the *Naval Magazine*, the *Sailors Magazine*, and the *Army and Navy Chronicle*. Non-naval journals also included a growing number of articles about the Navy or by the officers themselves. The naval publications of the 1830s offered junior officers the first opportunity (under pseudonyms) to engage publicly with senior officers and citizens on topics they had begun to write about in their own private correspondence and journals.

Periodicals

Whether in non-naval or Navy-specific journals, officers often wrote under pen names, a practice not strictly confined to naval officers. The British colonies and eventually the United States witnessed three generations of pseudonymous authors prior to the 1830s. Perhaps the most recognizable was a young Benjamin Franklin's use of Poor Richard. Later, during the debates surrounding the ratification of the Constitution in 1787 and 1788, the Federalist Papers had a collection of authors writing under the pen name Publius and Aristedes. Cato, Centinel, and Brutus wrote the Anti-Federalist Papers. The contentious nature of the constitutional debates warranted the use of pen names, which protected the authors' privacy but also focused the debates on the messages rather than the messengers.

The same was true of the public naval debates of the 1830s. In a variety of journals, essays appeared by authors using only a letter: C, D, M, and X. Others chose names or phrases such as Coquille, Neptune, Candor, and A Friend of the Navy. Publishing under their own names would have proven detrimental to their careers, and the public would have dismissed their concepts as those of inexperienced, junior officers who lacked the maturity of the elder captains. As articles were written anonymously, the public focused on the content and the messages rather than potentially critiquing a known author and their motivation. Thus, they could gain a wider audience. Pseudonyms were an effective means of challenging conventional thinking, but not all officers agreed with the method. One officer wrote of the "odious system of anonymous communications. Fear of power originated it; fear of detection has continued it. . . . In the warfare of anonymous communications, we pierce the masque, without thought of the wound we inflict on him whom it conceals, but shelters not."[28] By contrast, it took until 1912 before the Royal Navy found an anonymous forum for debate with the *Naval Review*.

Like the British *United Service Journal* (1829–41), professional magazines emerged in the 1830s to provide a greater awareness for both military services with some focusing strictly on the Navy. All played a role in initially advancing naval thought. The first superintendent of the Naval Academy,

Franklin Buchanan, thought the *Military and Naval Magazine* so important that he recommended to Secretary of the Navy George Bancroft that it become part of the naval school's library. Its editor, Benjamin Homans, must have known that officers were keeping private journals and issued a call for stories and anecdotes of their "piquant adventures."[29] In the inaugural issue, one officer wrote of the brotherhood of the two services and hoped that officers would help make the magazine "the repository of American talent, and the altar on which all our literary oblations shall be offered."[30] The magazine survived only three years. Homans attempted a similar format but as a weekly periodical in 1835 that completed its run in 1842. It was, as he wrote in the inaugural issue, a "means of obtaining information that will be highly valuable to the officer and to the citizen."[31] The *Sailor's Magazine and Naval Journal* primarily found a civil maritime audience; it, too, recognized the growing abundance of materials for such a periodical. "The mine" wrote the editor in August 1829, "has, in fact, but just begun to be opened."[32]

Of the new periodicals, the most important was the *Naval Magazine*, the publication of the Naval Lyceum whose purpose was to solicit from all fellow officers "such original papers . . . on all subjects, directly or collaterally, connected with the elucidation and diffusion of Nautical and General Science, and professional knowledge."[33] The *Naval Magazine* found the perfect editor with Reverend Charles S. Stewart (no relation to Commodore Charles Stewart), a Navy chaplain, missionary, and well-known diarist who had published his journals about his round-the-world travels aboard Navy ships. Although it was a short-lived periodical, the magazine "served as a catalyst for a budding movement seeking to institute a number of reforms and technological advances in the Navy."[34] The Naval Lyceum specifically aimed at stimulating its members intellectually and exciting the esprit de corps of the service. Its journal was a natural extension of that intent. It provided the public forum for the next generation of naval officers. Junior officers saw opportunities for naval evolution in terms of personnel reform, platform innovation, and policy changes. The journals provided them the outlet for their frustration and desperation to ensure that conventional thinking did not impede that change. The pseudonymous Sinclair wrote: "The spirit of the times and the necessities of the Navy loudly declare that change is requisite. We cannot remain as we are; for things are fast tending to that point, when the prospects of a vast majority of the officers of our Navy, will be desperate in the extreme."[35]

The periodicals offered a variety of essays. These included biographical essays, articles on emerging technologies, travel notes about the ports and countries the officers visited, assessments of foreign naval capabilities, naval policies, and ideas about how to improve the Navy. Some of the topics were extremely contentious, particularly when senior officers weighed in on the debate. These debates were not confined to the pages of the military periodicals. The newspaper *Baltimore American*, for example, noted the ongoing discussion in the *Army and Navy Chronicle* about the creation of the grade of admiral in the Navy.[36] The most common themes involved steam power, naval ranks, a naval school, and other navies and countries.

Just as with their letters, junior officers wrote articles for the journals, sharing their experiences and outward-looking views on other countries and navies. Although the preceding generation of naval officers had traveled the globe, they published few articles, especially since the various media were minimal. The *Analectic Magazine* (1813–20) was the only relevant medium; this magazine included some American stories and biographies of naval officers, but most essays were reprints from British periodicals. The articles in the 1830s, by contrast, offer a view into the maturity of the junior officers because of how they observed the world and the trends and challenges they offered.

Officers began to think strategically. The *Military and Naval Magazine*, for example, included a review of Carl von Clausewitz's *On War*. Young officers developed a more nuanced and progressive understanding of naval strategy and war fighting. Author D wrote, "'In peace, prepare for war' is the principle upon which our Navy has been sustained."[37] One journal reprinted Charles Dupin's "Naval Force of Great Britain" in which he wrote, "The absolute force of navies ought not to be measured by the number killed and wounded, in combats between [the French and British] navies."[38] Although the United States had been at peace with Britain for more than a decade, young officers continued to view the empire as the greatest threat but also as one from which they could continue to learn. The same was true of the French. Other issues included a translation of the article "Naval Tactics for the French Navy," an article about the size, composition, and strength of the French and other navies, and "Improvements in Shipbuilding" about British ships. In a different case, an anonymous lieutenant discussed the conflict he saw in Ecuador, its causes, and the failures of its leadership.[39]

Officers in the 1830s began to discuss options for naval schooling. One anonymous junior officer admitted that this subject had been on his mind throughout his naval service and offered his essay with the specific intent that the Department of the Navy consider it for petty officers. Petty officers had served on US Navy ships but they did not have a uniform or insignia until 1841. In addition, their ship's captain appointed them, and, thus, they served only while the captain was on board.[40] A school on a ship educated apprentices aged thirteen to fifteen years old. "The Navy," a junior officer argued, "will soon become popular with that class of the community we must always depend upon for seamen . . . [and] inducements should be held out to them to remain in the service after they have served their apprenticeship."[41] This proposal represented a significant shift.

It was, however, the naval school for officers that captured the imagination of the authors. A naval school represented an absolute necessity to properly teach science. Moreover, schools at sea were not conducive to learning. "A place cannot well be found more unfavorable to habits of close application, than a man-of-war."[42] In this new age, specialized professors of the sciences were a major requirement, and it would be costly to have one or more on ships; rather, they ought to be centrally located on shore at an academy. The naval school should rest on the example of the experience of the Military Academy at West Point: "If it be proper to fit men for military command at home, it is certainly not less so to prepare for the naval profession those who are to represent our country in distant climes."[43] A naval academy could also serve to vet applicants and candidates. "The first examination for admission," wrote A.S.—likely Alexander Slidell Mackenzie—"would reject many applicants, and the subsequent years of probation would winnow away all the chaff, all the incorrigibly stupid, all the vicious, all the insubordinate."[44] A.S. also suggested a "school ship" for midshipmen. By coincidence, just five years later, Mackenzie commanded USS *Somers*. Without a trial, Mackenzie executed a midshipman (the son of the sitting Secretary of War) and two sailors for suspected intent to mutiny. His heavy-handed and, arguably, illegal action in responding to growing evidence aboard was one of the factors that contributed to the creation in 1845 of the permanent naval academy, which could teach officers about ethics among other courses for the naval profession.

No issue was as contentious in the 1830s military journals as the creation of the rank of admiral. John Paul Jones had first proposed the rank of admiral during the American Revolution, in the hope of being named so himself. Absent that opportunity, Jones pursued a flag officer billet and served in the Russian Navy. Fifty years later, the Navy was full of senior captains. By 1835, the average length of service of captains was 33.2 years; their average time in grade was 14.7 years.[45] Lieutenants faced a similar seniority problem. Without the retirement of captains, and with no rank between captain and lieutenant, lieutenants might wait decades for a promotion. Junior officers vocalized their objection to the current system through the military journals. They argued that the Navy ought to dismiss such officers from the service if they could not perform their duties because of age or disability. Moreover, the Navy should terminate commissions based on political influence. Ship command based only on years in service should rest on intelligence, professional skills, and habits.[46] Instead of admirals, they presciently suggested a new ranking structure of commanders, lieutenant commanders, lieutenants, second lieutenants, and midshipmen—a structure that became the standard later in the nineteenth century and largely continues in the twenty-first-century US Navy.

The proponents of the rank of admiral were haphazard in their zeal. "Let Congress create as many admirals," Sinclair wrote, "as in their wisdom they may deem necessary."[47] Alexander Slidell Mackenzie stated unequivocally that the Navy needed admirals, "not merely as a stimulant and reward for faithful service, but as an all-important element of discipline." He recommended one admiral, four vice admirals, and ten rear admirals—fifteen total for a Navy with barely double that number of ships. Proponents used the argument of discipline, but it was far more about prestige manifested by more senior ranks. Such a formidable Navy required more than a captain to command it. C lamented that if the ship of the line *Pennsylvania* (which, at 120 guns, was one of the largest in the world) "go abroad with a pendant at her mast-head[,] . . . let her be the first to show the flag of an American Admiral."[48] The younger officers suggested that the editors tended "too much to ingratiate with the Captains, in the matter of Admirals, to which the interests of the younger officers are opposed, and the creation of them [is] only advocated by the older Captains and their friends."[49] It was not the last time in US history that junior officers believed publications were the tools of the senior officers.

Concepts promoted by naval officers were insufficient to effect change. Other naval officers became aware of the ideas, as did policy makers; however, elected officials required a greater inducement for action. That could only come from constituents, and constituents are influenced by popular culture. In the 1830s, the most common popular medium, literature, promulgated ideas in popular culture. Literary figures from the period such as James Fenimore Cooper, Richard Henry Dana, and the entire Knickerbocker Group (Washington Irving, James K. Paulding, etc.) shared a passion for the maritime and naval environments. They provided a unique tie between the Navy and its vision that the broader, increasingly literate American public articulated and popularized.

This was the first period of truly unique American writing that appealed not only to a domestic audience but a foreign market clamoring for anything new from the former colonies. The new American literary collective had its genesis during the War of 1812 with the *Analectic Magazine*, published from 1813 to 1820. Comprised mostly of newspaper articles collected from the United States and Britain, Irving edited it during the war. Two decades later, he became of one of the most popular literary figures in America and served in the Jackson administration. He was editor when Francis Scott Key published "The Defense of Fort McHenry," which was later set to music as "The Star-Spangled Banner," later America's national anthem. Irving's personal friendships with naval officers like Stephen Decatur influenced his own biographical sketches of war heroes James Lawrence, Oliver Hazard Perry, and David Porter and produced the "most sophisticated naval writing of the war."[50] Author James Kirke Paulding thought the essays in the *Analectic* so important for officers that he recommended to Secretary of the Navy Samuel Southard that he include them in the library of every ship.[51]

Journals that were not Navy-centric were just as important as their maritime counterparts since they broadened the potential readership from those who were in or cared about the Navy to those who should care about it. These journals were part literary and part news with some biographical sketches. Just as the American art world experienced its own distinctiveness with the Hudson River School in the 1820s and 1830s, a similar group emerged from Tarrytown, New York. The Knickerbocker Group had ties to the Navy and advocated its role, even subtly, in national security as well as the importance of a maritime component of American society. The *Knickerbocker Magazine*, founded in 1833, included some of the most prominent literary figures of the period. James Kirke Paulding later served as secretary of the Navy under Martin Van Buren. Henry Wadsworth Longfellow's namesake was a naval officer killed at Tripoli. Oliver Wendell Holmes's poem "Old Ironsides" would later save the USS *Constitution* from being broken up.[52] James Fenimore Cooper had served as midshipman prior to his literary career. In addition to penning the first history of the Navy, he also wrote a series of articles on naval policies in the 1830s. The *Knickerbocker* published a piece by Jeremiah Reynolds titled "Mocha Dick: The White Whale of the Pacific," which became the foundation for Herman Melville's more famous book *Moby-Dick*. The *Southern Literary Messenger*, founded in 1834 in Richmond, Virginia, and edited by Edgar Allan Poe, provided a favorable review of Lieutenant Matthew Maury's book on navigation. It included essays about the Navy's first round-the-world voyages and an extensive series on naval policies by junior officers. The relationships between naval officers and literary figures continued through the next decade, such as the growing connections between James Fenimore Cooper and Matthew C. Perry.[53]

Another member, Washington Irving, was the son of a Royal Navy petty officer. In 1828 he published the *Life and Voyages of Christopher Columbus* and followed it up with a volume in 1831 on the *Voyages and Discoveries of the*

Companions of Columbus. Irving was an early critic of the militia and volunteer units as "miserable military quacks" during the War of 1812, but recognized the value of the Navy. "We are all live, at present," he wrote in late 1812, "in consequence of our naval victories. God knows they were well-timed to save the national spirit from being depressed."[54] While traveling in England in the 1830s, he maintained his relationships with naval officers (for example, his nephew Edgar Irving, a former midshipman and later Marine Corps officer, was a member of the Naval Lyceum), while trying to help Alexander Slidell Mackenzie publish one of his own works.

When Irving left the United States for Europe after the War of 1812, the country was largely rural with a capitol burned by the British in retaliation for the attack on York (now Toronto), Canada. He returned seventeen years later to a nation of railroads, steamboats, and a populist president. Writing to his friend Alexander Hill Everett in 1829, Irving averred that he had no political ambition after being offered the position of secretary of legation in London.[55] He planned on being as clear as possible from party politics. Whether it was the eventual allure of politics or the need for steady income, he finally accepted the position as secretary of legation.[56] He was soon criticizing the cabinet and the media in support of Jackson, his benefactor. When he was in Washington, the president received him in the White House.[57] It was the first of several recorded meetings between the author and the president.

James Kirke Paulding and Congressman Gouverneur Kemble later recommended Irving for the position of secretary of the Navy, but Irving refused. "I shrink from the harsh cares and turmoils [*sic*] of public and political life," he wrote, "and feel that I am too restive to endure the bitter hostility, and the slanders and misrepresentations of the press, which beset high station in this country."[58]

Other members of the Knickerbockers were less involved but nevertheless had ties to the Navy and to maritime industry. William Cullen Bryant, for example, a poet and editor in chief of the *New York Evening Post* from 1828 to 1878, was a close friend of Richard Henry Dana, author of *Two Years Before the Mast*. Others had similar ties to Dana, Nathaniel Hawthorne, and Henry Wadsworth Longfellow. Paulding was already an established figure in the Navy as well as for his literary work. The son of a ship captain, he served as secretary for the Navy's Board of Navy Commissioners (1815–23) and as a Navy agent in New York City (1824–38). In the latter capacity, he identified shortcomings in the libraries of Navy ships. Most of the books he recommended were published within the previous three years. They covered naval gunnery, naval architecture, atlases, and other practical books. Included in his recommendations was Irving's *Life and Voyages of Columbus* because he knew of no work worthier of being placed in the hands of Navy officers, "or one more calculated to inspire them with noble ambition."[59]

Throughout his tenure in New York, Paulding wrote in private and public of his views on the concepts of self-government, monopolies, and the works of other authors like the first volume of US history by George Bancroft, who, like Paulding, later served as secretary of the Navy. Heavy on the minds of these literary figures was the threat of war with either France or Britain. "Everybody here is talking of war," Paulding wrote to Martin Van Buren, "and my old friend Fennimore [*sic*] Cooper is in a most belligerent attitude every day in Wall Street."[60] Paulding recognized in France's case that it was not seizing ships and impressing seamen as with the British two decades before.

In 1838, President Martin Van Buren named Paulding secretary of the Navy. Irving often wrote him for favors—recommending the son of a New York politician for the position of artist for the South Seas Exploring Expedition, suggesting officers for promotion, and requesting a commission for his brother. Paulding's frustrations with the administrative and political components of the position could not have been clearer when he wrote to Irving that "the members of congress too, have got a habit of eternally interfering with appointments and destinations."[61] Yet, Paulding was not averse to supporting his fellow Knickerbockers. As secretary of the Navy, "as a brother Scribe," he directed James Fenimore Cooper's naval fiction and history books "to be added to the Libraries of the Public Ships, for I know not where our young officers can find better practical illustrations of Seamanship than they contain."[62]

Although Paulding became a Navy secretary, no Knickerbocker had as much knowledge or impact on the Navy as James Fenimore Cooper (fig. 2.3). A midshipman before the War of 1812, Cooper formed lifelong relationships with the naval officers with whom he had served such as William Shubrick, Isaac Chauncey, and Richard Dale. In the pages of the *Naval Magazine*, Cooper expressed his thoughts on ranks, manning, and ships. He favored increasing taxes to increase the size of the Navy by 50 percent. His personal correspondence with Shubrick specifically about the ship of the line *Pennsylvania* under construction in 1836 to 1837 demonstrated an intimate knowledge of its lines and potential uses.

While he continued to write fiction, including maritime-themed works, Cooper set out to write the first major history of the Navy. Long-time chief clerk to the Department of the Navy Charles W. Goldsborough wrote the *United States Naval Chronicle* (1824), a brief history through 1801.[63] Cooper's two-volume work, by contrast, was the first American naval history book to

2.3. James Fenimore Cooper, c. 1861. Though better known for his novels, James Fenimore Cooper was a prolific writer on naval affairs in the 1830s. Library of Congress, Prints and Photographs Division.

have a unifying theme—the high character of the American naval officer.[64] The book received mixed reviews, primarily because of his treatment of Oliver Hazard Perry and Jesse Duncan Elliott during the Battle of Lake Erie, namely the belief that Elliott failed to engage his ship with the British while Perry's ship bore the brunt of the enemy's gunfire. The *Army and Navy Chronicle* wrote that Cooper's work was a much-needed history and that surpassed Cooper's "power of nautical description."[65] A British review criticized him for "detracting from the fame of the British Navy," a task, it argued, that was clearly beyond his powers based only on loose generalities.[66] Whether or not he was correct in his assessments or writings, Cooper's body of work—both fiction and nonfiction—was a major contribution to the public's awareness of the Navy at a period of major change.

Nor was literature the only medium that enabled the public to learn about the sea services. America began to recognize its naval heritage through sculptures. The Tripoli Monument, dedicated to the dead naval personnel during the Barbary War, was erected at the Washington Navy Yard and moved to the Capitol in 1831 and then to the Naval Academy in 1860. The intent of the Tripoli Monument was to memorialize the sacrifice of the Barbary War sailors and inspire the public; the mission then changed to that of inspiring the students of a new school.[67]

Education: Creating Noblemen in the Defense of the Republic

By 1840, more than two million American youth had enrolled in schools, including in the growing number of colleges.[68] Education was part of the wave of democratization with the election of populist Andrew Jackson in 1828. It also reflected the first wave of immigration by injecting previously unschooled youth, the second revival period for American religion, and the introduction of new technologies that spurred the Industrial Revolution. These forces led to a shift in the country's approach in how to develop its youth. The same debates in the civilian sector applied to how the Navy modified its own education of officers.

Education in the early republic varied because of federalism, geography, and population. The nature of the federalist government separated powers between national, state, and local governments; the Constitution did not include education. Therefore, it fell to state and local governments. Diversity within each state also challenged how policy makers treated education. In New York, for example, only a quarter of the state's youth were enrolled in schools. Some states required towns to build elementary schools based on population. The United States only had 5.3 million residents (including 900,000 slaves) in 1801, but its population would quadruple by 1850. Britain, by comparison, had a population of 9 million in 1801 that would double by 1850. In 1800, only six US cities had a population of 10,000 or more; by 1830, twenty-three cities had 10,000 or more. Urbanized areas, especially in the northeast, had the population and tax base to support school systems. Most midshipmen were from urban environments and consequently already had some exposure to school by the time they were commissioned.

One historian notes that "after 1830 a new generation of educational reformers appeared, and the tide began to turn."[69] Massachusetts reformer Horace Mann, brother-in-law of literary figure Nathaniel Hawthorne, became the father of the common school movement and founded the *Common School Journal* in 1838.[70] Mann and other reformers believed education provided the appropriate discipline in creating responsible citizens for the republic.

The Prussian model influenced American education and its reformers more than British education despite the latter's Anglican school establishment. The Prussian system emphasized discipline and, more importantly, certified and trained teachers. Reports on Prussian education began to reach the reformers including Mann and his fellow Massachusetts educator George Bancroft, who founded the US Naval Academy, before he served as the secretary of the Navy. Like its Prussian model, American education found itself also driven by nationalism. Common schoolbooks began to appear in the 1830s. Noah Webster, for example, designed textbooks "to promote a consistent American language and a common knowledge of geography across regional boundaries."[71]

A map of the United States, often copied, dominates another common schoolbook, *Rudiments of National Knowledge, Presented to the Youth of the United States, and to Enquiring Foreigners*.[72] Superimposed on the map is an eagle in full flight, its head dominating New England, its wings effortlessly spreading across the Midwest and beyond the established borders and its talons firmly planted in the region of Florida, still controlled by Spain and populated by Native American tribes. The eagle's eyes fix on the Atlantic and beyond. Was the image an accident or intentional? Perhaps it was an indication that the publishers wanted American youth to recognize that the country would continue to move west, that conflict in Florida was likely, and that it would defend its territory from European powers.

In contrast, British naval education rested on the needs of empire and the idea that only a select few—primarily from the aristocratic class—would lead their ships and protect their civilization. By 1700, there were three methods of becoming a commissioned officer—as an apprentice to a senior officer with each officer allotted a certain number, as a nominee of the admiralty, and as a rating.[73] Like the early US Navy, instructors at sea were mostly clergymen. Eventually the British instituted schoolmasters, with ninety-eight appointed between 1806 and 1824 but only twenty serving regularly in the fleet.[74] By the 1830s, the Royal Navy had abolished schoolmasters.

European powers established naval schools as the wars of the seventeenth and eighteenth centuries developed (France, 1682; Denmark, 1701; Russia, 1715). The French also later established the Ecole Navale in 1830 on the ship *Orion*, the first of the formal naval academies. Britain had established the Portsmouth Naval Academy in 1733, but few entered it because it threatened the patronage system of the senior ranks. Arguably, the patronage system in the Royal Navy largely worked; no senior officer wanted to gain a reputation for advancing less capable officers. The academy closed in 1806. Two years later, the British founded the Royal Naval College, and a Royal Naval Academy opened from 1833 to 1837. The British would create a formal naval training school in 1850 on board HMS *Britannia*. The admiralty's sentiment was no different from that of its American peers. There was a dependency on tradition and resistance to cultural and, therefore, naval educational change.[75] King William IV, the "Sailor King," articulated this resistance in his comment that "there was no place superior to the quarterdeck of a British man of war for the education of a gentleman."[76] This was similar to Commodore Stewart's quote about American naval education, that the best school for the instruction of youth in the (naval) profession is the deck of a ship.[77]

However, the Age of Enlightenment introduced new disciplines and applications of scientific theory, particularly with the advent of steam warships and new ordnance. The US Navy needed a new generation whose curiosity could harness new technologies. More importantly, the Navy needed to properly educate them to employ these technologies. The creation of West Point, for example, aimed to meet the needs of a nation desperate to defend itself from foreign incursion. Part of that need was to train engineers on warfare such as the use of artillery or the construction of fortifications.[78] The unique nature of the sea required that midshipmen learn their trade on the oceans during the Age of Sail, but as warfare and platforms evolved, the Navy and the nation required a more formal school.

Books became the junior officers' primary opportunity to learn about theory and history on ships. What they read provides a glimpse into their awareness of naval history, science, nature, and politics around the world. By 1839, the Navy furnished a variety of books to ships' libraries[79]—histories of Greece, Rome, and England; Vattel's *Law of Nations* and Jacobson's *Sea Laws*; treaties with foreign powers; descriptions of the voyages of Columbus, Drake, Magellan, Cavendish, and the USS *Potomac*; and others. Midshipman Thomas Dornin wrote about some of the books available on the *Falmouth*: "I have been lately engaged in perusing the lives of Collingwood and Nelson; what a different feeling you experience from reading these two works. The spirit nobly displays unshaken valour, keen diplomatical [*sic*] knowledge and a mild Christian like benevolence and a wonderfully just estimate of human character. It should be perused by every Naval Officer and would be no contemptible addition to the library of any statesman. But I felt a deep and sensible pain on reading Southey's life of Nelson, which cannot be deemed otherwise than impartial and just."[80] At this time, biographies of American naval officers were rare, hence the reliance on books about British officers. James Kirke Paulding made books for adoption fleet-wide in 1838. Four years later, a formal printed list of books for ships appeared in Navy regulations. Midshipmen, however, still needed education.

Shipboard schooling was a natural but inconsistent means of educating midshipmen. In 1813, schoolmasters served on ships that had at least twelve midshipmen. Until 1815, this meant only frigates, since the Navy's first ships of the line were not completed until after the war. On smaller ships, such as brigs, sloops, and schooners, the schoolmaster role was collateral duty for chaplains. By the 1820s, the Navy required basic qualifications for schoolmasters, but individual captains, not the Navy, selected the clergy that served on their ships. Consequently, their ability to teach scientific subjects remained limited or at best inconsistent from ship to ship. In the 1830s, the Navy attempted to resolve this discrepancy.

The Navy recognized that during a peacetime environment, shore-based schools could educate midshipmen more efficiently before they deployed on their ships. The Navy hired qualified instructors and based them at the three major Navy yards in Boston, New York, and Norfolk. Midshipman Charles Steedman reported to the naval school at the New York Navy Yard in 1828 under the command of Commodore Isaac Chauncey and Captain Francis Gregory. The school included thirty to forty midshipmen where Edward C. Ward served as professor of mathematics, and Professor Moulle taught Spanish and French. By Steedman's calculation, before the establishment of the Naval Academy, nearly half the midshipmen came from appointments by politicians from Virginia, Maryland, and the District of Columbia.[81]

In some cases, the Navy and the private sector attempted innovative and collaborative measures. For example, in 1830, the Columbia College's (later University) new president William Duer offered to Commodore Isaac Chauncey schooling for the local young officers.[82] On January 16, 1830, Duer established a scientific and literary course open to both matriculated and nonmatriculated students. The course was like those already offered by Columbia with the exception of replacing Greek and Latin languages with modern languages.

Chauncey was cautiously intrigued by the offer and wrote to Secretary of the Navy Branch: "This proposal is a liberal one, not more expensive than the Navy yard schools. . . . I certainly should prefer a Naval School, if Congress would authorize one." Although, for unknown reasons, he found it impracticable at that time, he was amenable to the possibility. Until that could happen, Duer also recommended attaching a naval officer to the college for "superintending the young officers and enforcing discipline." Within a month, Chauncey identified the challenges and approved a plan for the education of naval officers.[83] It is unknown if this collaboration between the Navy and Columbia resulted in any Navy students or graduates, although there is some evidence of at least one participant.[84] What is known is that Duer had come far since his days as a midshipman three decades before during the Quasi-War. Threatened with a court-martial for stabbing and threatening to murder another midshipman and striking a lieutenant, Duer relied on his family to save him. His mother was Lady Catherine Duer, whose father was Lord Stirling (William Alexander), an American Revolutionary War general. She appealed to President John Adams to allow him to resign. Adams turned the matter over to Secretary of the Navy Benjamin Stoddert, who blamed the ship's captain for allowing too much wine in the wardroom. Duer resigned, attended law school, was elected to the New York State Assembly, and became a judge before assuming his role at Columbia.[85]

Few institutions—except for lyceums—were better manifestations of the Age of Enlightenment than schools, representing as they did the progress of human knowledge and mind. The lyceum movement in the United States offered adults an opportunity to learn through public lectures as noted figures traveled the country from lyceum to lyceum. The first lyceum was established in 1826 in Massachusetts.[86] The Brooklyn Navy Yard lyceum would serve as a lecture hall, museum, library, and post office for naval officers. In the absence of a formal, national naval school, one junior officer asked in 1834, "Are we always to remain stationary, while the arts, sciences, and a general knowledge are in a course of rapid progression and diffusion?"[87] The Navy traversed the globe requiring, by its nature, cooperation with foreign entities for supplies. Its officers interacted with other navies, government officials, and, just as importantly, individuals with commercial interests. Thus, the Navy needed educated gentlemen to carry out the affairs of state and industry. Common schools and the naval schools were insufficient to meet that need, particularly in the areas of developing moral character. The Naval Academy, while proposed for nearly fifty years, might not have come into existence without a fundamental change in platforms and technologies and junior officers who recognized and advocated for this change.

As early as the American Revolution, John Paul Jones had recommended the foundation of a naval academy—or rather a series of academies at various bases. In 1806, Captain Thomas Truxtun "lamented the lack of naval education . . . especially when it came to the more theoretical aspects of the profession."[88] Truxtun also bemoaned the lack of naval pride at the time, even in the wake of the First Barbary War, though this national attitude changed with the War of 1812. In 1815, the Board of Navy Commissioners report also supported the creation of a naval academy. Secretary of the Navy Southard suggested that a naval academy would "improve discipline and reduce the number of courts-martial."[89]

Supporters and opponents both invoked naval history to support their case for a naval academy. One congressman argued that it would "give to all the young men in our Navy the means of becoming Decaturs and Perrys."[90] The congressman failed to recognize that neither Decatur nor Perry were products of a naval school, a case made by one anonymous essayist who asked, "Could a Rodney, Nelson, or a Collingwood, have been made in such a school? If this system is continued shall we not in vain look for future Decaturs, Lawrences, Perrys, and MacDonoughs?"[91]

The Crucible of Enlightenment: The Naval Lyceum

The nature of a Navy is that the entire fleet is rarely, if ever, at shore simultaneously. Some ships are deployed, while some are in their home nation undergoing maintenance and repairs. Even ships that are ashore are scattered across different ports, assuming a nation has more than one port. During the 1830s, for example, the Navy had seven homeports.[92] At any given time, one-third to one-half of its ships were deployed around the globe. Thus, in the Republic of Ideas, correspondence and publications in various periodicals were important components in disseminating original and evolving concepts. These provided a valuable link among the new generation of naval officers at home and in foreign lands, but they lacked a critical element to unite them for a common cause—a home. In this case, the home became a structure and accompanying organization that enabled open-minded senior officers to encourage ideas of the junior officers. The Naval Lyceum provided that home, a crucible for naval reformers that provided legitimacy through a permanent, professional organization that could advance new concepts. The Naval Lyceum was not the first attempt at a professional military organization in the United States. As the nation concluded the First Barbary War, returning officers established the United States Military Philosophical Society. Its purpose was to collect and preserve the military science from veterans of the American Revolution and share scientific knowledge about warfare. During its existence from 1806 to 1813, the society's members included the sitting president and secretary of war, elected officials, and Army and Navy officers.[93]

The lyceum movement of the early nineteenth century sought to inform and educate adults through lectures and other public events.[94] Massachusetts established the first lyceum in 1826. Other lyceums quickly followed in New England and then the mid-Atlantic states and eventually some southern metropolitan areas.

In December 1833, officers stationed at the Brooklyn Navy Yard made a public announcement about the founding of the Naval Lyceum, the purpose of which was to "elevate and adorn the character of our Navy, by placing within the grasp of its officers the means of acquiring professional and general information . . . to stimulate the members of the profession, by creating a common interest in the result, to new energy in the steady and zealous pursuit of knowledge, as the grand source of moral power, and to bind yet more closely the ties which unite them, but erecting a National Society."[95]

The officers advised the Board of Naval Commissioners on the establishment of the organization in which they proposed a cabinet of natural and artificial curiosities and specimens, as well as the creation of a library, to be built with funds from membership fees. They sought to stimulate the zeal, talents, and enterprise of officers to collect and furnish specimens collected from abroad.[96] Soon officers and other members began donating works. Josiah Tattnall, later a senior officer in the Confederate Navy, provided a copy of a *History of Naval Architecture*; James Fenimore Cooper donated some of his own work. By 1835, the library consisted of 1,134 volumes, 141 European periodicals, 148 American periodicals, and 14 books of charts and maps.

The Lyceum published its own periodical, the *Naval Magazine*, for two years (1835–36), but other periodicals echoed this curiosity about the world. The *Military and Naval Magazine* issued a call to its readership: "We must look for collections of antiques—of nautical surveys—for descriptions of foreign people and countries; their manners, commerce, and institutions; their climate, soil and productions."[97] One Mediterranean-based officer in 1836 noted as he passed Ithaca that it was "the birthplace of Telemachus of which Ulysses was King" and cited Lord Byron's works as his squadron formed north of the Peloponnese before visiting the Acropolis. Byron had, in fact, visited the USS *Constitution* when it was in the Mediterranean.[98] The Naval Lyceum was soon filled by artifacts from around the globe. These included antiquities from Egypt and Greece, wood sculptures and weapons from South America and the Pacific Islands, and items from East Asia. As a result, the Lyceum became one of the first repositories in the United States a decade before the Smithsonian Institution.

This inquisitiveness was an indication that the Navy and the nation were not insular. The Naval Lyceum was the foundation organization for the officers to learn about that world. The Navy's officers were outward looking and were forming an understanding of foreign cultures, histories, and geography. It was also an indication that the Navy must evolve and take an increasing role in world affairs. One historian states that Matthew Perry (fig. 2.4) was the "driving force behind [the Lyceum's] organization and activities."[99] While this is largely true, Perry also needed support from the new generation of naval officers eager to challenge conventional thinking and embrace new concepts through an institution of professional learning like the Lyceum. Perry served as the first vice president with Commodore Charles Ridgely serving as the organization's first president. The second vice president was Tunis Craven, a naval storekeeper and son-in-law of Captain Thomas Tingey, who had been the long-time commander of the Washington Navy Yard. Tingey was also a member of the Washington-based Columbian Institute for Promotion of the Arts and Sciences.[100] The institute was responsible for the creation of the US Botanical Gardens on Capitol Hill. It is possible that Perry and the other officers drew from the mission of the institute for the Lyceum. Lieutenant William C. Hudson served as the Lyceum's corresponding secretary. In 1838, he received command of *Peacock* and placed second in overall command of the South Seas Exploring Expedition. In 1857, he was the first commanding officer to attempt to lay a transatlantic telegraph cable in cooperation with HMS *Agamemnon*, a feat they accomplished the following year.

The meetings began in Perry's office at noon every Monday. The first year, the Lyceum had eighty-eight New York resident members, eighty-three nonresident members, and fifty-two corresponding members including four British captains, Prince Charles Bonaparte, and ship designer Samuel Humphreys, son of Joshua Humphreys, who designed the first six frigates of the Navy. Honorary members included President Jackson, Martin Van Buren, former presidents Madison and John Quincy Adams, Joseph Bonaparte, and

2.4. Commodore Matthew Perry. Master Commandant Matthew C. Perry was instrumental in guiding the navy's enlightenment during Jackson's presidency through the establishment of the Naval Lyceum. Courtesy of the US Naval Academy Museum.

literary figures such as Cooper and Irving. Conspicuous by their absence were some of the senior-most captains in the Navy, including Commodores James Barron and Charles Stewart.

Its membership reflected the interests of the new generation of naval officers (fig. 2.5). Of the 28 officers of the USS *Franklin* in 1817 still on active duty in the 1830s, all were members of the Lyceum, with the exception of Stewart. Of the 267 members during its initial 6 years, the near majority were junior officers—117 were lieutenants and another 48 were passed midshipmen. Fewer than half of the serving captains were members.[101]

Members also included twenty-nine Navy surgeons. Nonmembers were welcomed as well. Ships' wardrooms were invited to meet at the Lyceum when they pulled into New York.[102] Equally important were the numerous guests who visited the museum. The Lyceum's guestbook reveals a telling story about those whom the Lyceum's events attracted or whom it sought guidance from or, in some cases, whom to influence. One of the first guests was Richard Russell Waldron who a few years later served as purser and Navy agent with the South Seas Exploring Expedition. The Lyceum attracted academics like Spanish professor Miguel Cabrera; mathematics professor George W. Benedict of the University of Vermont, who later helped build a telegraph company; and Professor Benjamin Silliman, the first chemistry professor at Yale University. Guests included politicians like Congressman Robert Livingston and Senator Nathaniel Tallmadge as well as British, German, Mexican, and French military officers. Naval constructor John Lenthall frequented

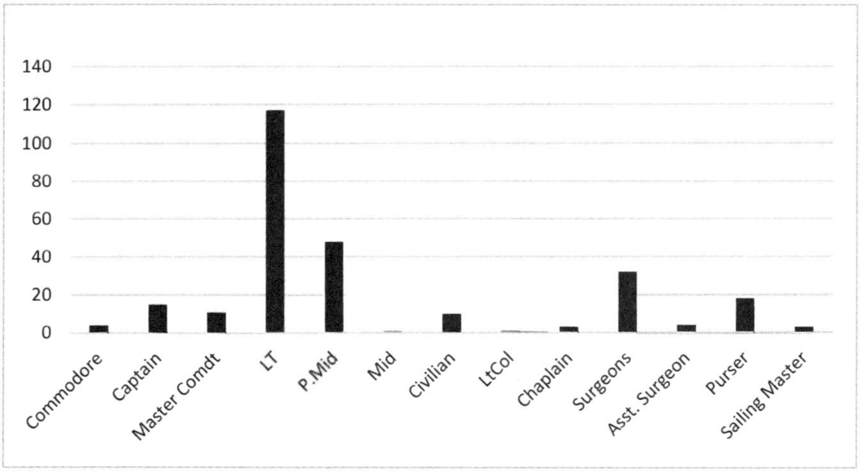

2.5. Membership of the Naval Lyceum, 1835–38, by rank.
Naval Lyceum account books, US Naval Academy Museum.

the Lyceum and was largely responsible for the ship designs used through the Civil War.[103]

The Naval Lyceum's existence helped to bond the officers and interested civilians during the 1830s about naval issues. First, this afforded the first opportunity for naval officers to gather and communicate ideas with one another since the Lyceum also served as a post office for their correspondence. Second, it provided a venue to meet others such as foreign military dignitaries; domestic Navy, Army, and revenue cutter officers; legislators; and academicians. Third, its existence led to other publications. In 1873, one of the junior officers and his son founded the US Naval Institute, which would publish the magazine *Proceedings* and the resulting concepts of American sea power advocated by Alfred Thayer Mahan and William B. Luce, both of whom had been members of the Lyceum and Naval Institute.

The same democratic wave that swept Jackson into power empowered a new generation of naval officers to challenge conventional naval authority. They were awoken by a new spirit and united by their ability to communicate with one another. In twenty-first-century parlance, they crowd-sourced their ideas such as investigating how new technologies could be used to advance the fleet, how they might achieve a common naval education, and the US Navy's expanding role for the nation.

While the officers were able to disseminate their ideas among each other and in publications, another factor played a significant and necessary role in advancing their ideas. This was the first decade in which the officers could share those ideas with like-minded public figures. The nation's early literary figures were instrumental in propagating their concepts. Luminaries such as James Fenimore Cooper and others sought to appeal to and convince the nation that it had a maritime destiny. Although they wrote nonfiction, their sea-based fiction was more popular and helped ingrain a sense of America's role on the high seas. These literary figures were able to move the issues and the romance of the sea and the Navy beyond the confines and echo chamber of the Department of the Navy. They engaged the American public by conveying ideas through naval-oriented fiction such as Joseph C. Hart's 1834 *Miriam Coffin*, in which he wrote that the country was "informed by a strident maritime nationalism. Supremacy on the ocean is America's great national destiny."[104]

Although they had the ideas, the functions had to be implemented by sympathetic policy makers who had the authority to make changes. During a unique era, no one held more power and ability to do so than the president himself and the administration he built to enable these changes.

3

Governing the Navy

Jackson's inaugural address only briefly mentioned the Navy; yet, it demonstrated an understanding and respect for the Navy and its role for the nation. As with the other departments under his executive branch, the Navy immediately fell under financial scrutiny. He assigned one of his loyalists, Amos Kendall, to investigate its accounts, not to punish the Navy he had inherited from his predecessor but to better finance it in future years and determine if there was any fraud.

To administer his Navy, Jackson selected three secretaries over the course of his administration, all of whom had served their states as governors and senators. Though they each lacked naval experience, their combination of executive experience and legislative work on Capitol Hill voting on Navy budgets made Jackson's choices wise ones.

Jackson's governance of the Navy benefited from his personal expectations of justice, demands for accountability, and preparation for what he envisioned as future operations. He believed the Navy was a vital arm of American commerce and its growing influence. Yet, his administration of the Navy—and of the whole of his government—was narrowed by his excessive misplaced sense of honor and justice, as well as his belief that his opinion ought never to be challenged. His volatility sometimes undermined the stability of national—and naval—governance.

The debate on the size and form of the Navy had diminished compared to the constitutional debate in 1787. However, the ideas about the purpose, size, and composition of the Navy continued and remains an ongoing discussion in national security circles even in the twenty-first century. The navalists—those who proposed a standing Navy operating at various global stations—dominated in the post–War of 1812 environment. Antinavalists were in the minority but were not silent during Jackson's administration.[1] The antinavalists supported a more state-based structure for a Navy that would empower individual states to compete on ship designs and needs. The southern states, advocates of a smaller government and decreased federal budgets as well as commercial markets for cotton, were especially aware of the cost of maintaining a Navy. This would play out during the debate on whether to build the ship of the line USS *Pennsylvania* in 1835.[2]

Despite the limited amount of attention paid to the Navy during his inaugural address, Jackson began his presidency with a basic but essential understanding of the need for the Navy. Jackson demonstrated this understanding and vision in four ways. First, he called for a gradual increase in the Navy, a decidedly navalist approach. No antinavalist would have suggested during peacetime that the Navy remain as large as it was, much less suggest that it grow. What did Jackson foresee that required this gradual increase? Of paramount importance was the protection of America's commerce, whose needs grew with the rapid rise in population. In the 1830s, for example, America had supplanted England as the largest whaling nation. In 1828, the gross tonnage of American merchant ships engaged in whaling was 54,000 gross tons (whalers were generally a few hundred tons). By 1837, that had increased to 129,000 gross tons. It would finally peak in 1857 when the gross tonnage was 195,000.[3] The total tonnage of US vessels cleared to leave port in 1828 was 897,000 (six times the number of cleared foreign vessels). In 1836, more than 1.3 million tons of US vessels were cleared to leave port (twice the number of foreign vessels).[4]

America's economy also grew with new markets such as China. Although the first US merchant ship left for China in 1785, American commercial activity in the region demanded the establishment of the last permanent squadron by the end of Jackson's administration. In the 1830s, the American merchant fleet was second in size only to that of England.[5] One indicator of the growth of the merchant fleet during this decade was the total gross tons of documented vessels. In 1830, the merchant fleet totaled 1.2 million gross tons; by 1837, that number had risen to 1.8 million gross tons. The growth was much higher since the 1830s had seen the advent of smaller, more efficient steamships. In 1828, steamships totaled 39,418 tons; by the end of Jackson's second term, that number had grown to 154,765 gross tons. With more ships plying American trade came the threats the country had always faced, such as piracy. The potential for conflict with other powers, particularly the British whom he had fought against twice, required a standing and growing Navy.

Second, Jackson recognized that the Navy should be more than a coastal defensive force. It needed to display the flag in foreign waters, if not as a global power then as a force with a global mission to patrol distant waters. Jackson continued the naval expectations of his predecessors—Madison, Monroe, and Adams—who had supported permanent overseas squadrons in the Caribbean, Pacific, Mediterranean, and elsewhere in the post–War of 1812 environment. Jackson made it clear that the Navy was on distant stations to stay.

Third, Jackson knew the power of the rising tide of national patriotic fervor within the young republic. His casual mention of the Navy's "fame in arms" recognized the crucial role the small Navy had played in the Quasi-War with France, the Barbary Wars, and the War of 1812.[6]

Though the Navy had lost ships, its single-ship victories against superior powers had been unexpected and gave hope to the nation. In case the country was to engage in naval conflict during his presidency, he committed himself to ensuring the fleet was ready to avoid a reversal of its maritime fortune.

Fourth, Jackson proposed "the introduction of progressive improvements in the discipline and science."[7] Along with the previous factors, this term would underscore yet another unexpected aspect of Jackson's Navy that was shaped by the previously mentioned new era of naval enlightenment—a professionalization of personnel and liberality of the era's burst of scientific activity.

Jackson immediately formed his government and submitted names to the Senate for confirmation. This was his official cabinet; newspapers would later coin the term "kitchen cabinet" for those official and unofficial advisors closest to him.[8] Martin Van Buren, his secretary of state, had organized the Democrats for Jackson in the 1828 campaign. For secretary of war, Jackson nominated John Eaton, whom he had served with in the War of 1812 and as US senators from Tennessee.

What would be termed "The Eaton Affair" would impact Jackson's governance including that of justice within the Navy. Attorney and marksman Charles Dickinson had made intemperate remarks about Rachel Jackson. This led to a duel in 1806, in which Jackson killed Dickinson. In Jackson's worldview, a wife's honor ought always to be defended. In many ways, it was the same as his position on national security—"he sees an offence . . . and he takes action."[9] So it was with Eaton's wife, Margaret.[10]

The then Margaret O'Neal worked in her father's hotel a few blocks from the White House. She was popular and warranted the attention of many sons and nephews of Washington's elites, a fact she fully offered later in life—"I had the attention of men, young and old, enough to turn a girl's head."[11] At the age of sixteen, she married US Navy purser John Timberlake, a man twenty years her senior.[12] The following year, when he was sworn in as a senator, John Eaton began lodging at Franklin House.[13] Eaton and Timberlake developed a friendship, and when the latter needed help to resolve a financial issue arising from his losses when the British captured the USS *President* in 1815, Eaton sponsored a bill in 1820 to relieve him.[14] The relief bill failed. Timberlake then served with Matthew Perry on the USS *Shark*.

In 1824, Purser Timberlake boarded the USS *Constitution* for duty in the Mediterranean and gave Eaton power of attorney for the purchase and transfer of Franklin House.[15] Timberlake died on board on April 2, 1828, but accounts differ on the cause. Some suggested it was pulmonary disease, while others suggested he died by suicide. In either case, John Eaton took the widowed Margaret O'Neal Timberlake as his bride on January 1, 1829, two months before Jackson was inaugurated. As Eaton was nominated and confirmed as secretary of war, rumors about the couple were already swirling in Washington. Cabinet members, the wives of cabinet members, and Washington's social elites shunned Margaret Eaton, whom they believed was an adulteress. Members of Jackson's own family, Andrew and Emily Donelson, ostracized her.[16] As a result, Emily no longer served as First Lady and was replaced by Jackson's daughter-in-law, Sarah Yorke Jackson.

Coincidentally, Jackson's request for an overall accounting of the Navy naturally included an investigation into John Timberlake's finances while he served as purser aboard the USS *Constitution*. The investigation coincided with the spurning of Margaret Eaton, creating a perfect storm for Jackson. He referred to opponents of Eaton as creating an "unholy wicked & unjust conspiracy" and as "political combinations of Slanderers."[17] Jackson's rage over the affair seeps into his correspondence throughout 1830 and 1831. Jackson's first step in response to this perceived injustice was to replace the cabinet. On April 18, 1831, Martin Van Buren and John Eaton submitted their resignations. The following day, Jackson resolved to reorganize his cabinet, "this being the only course, looking to justice to all" and asked Secretary of the Navy John Branch (fig. 3.1) to resign.[18]

Branch resigned effective May 12, 1831. Secretary of State Martin Van Buren left May 23. Secretary of the Treasury Samuel Ingham, Secretary of War Eaton, and Attorney General John Berrien all left their positions in the middle of June. Only Postmaster General William Barry remained. Edward Livingston replaced Van Buren as secretary of state. Louis McLane was recalled from his post as minister to England to serve as secretary of the Treasury. Lewis Cass, governor of the Michigan Territory since 1813, replaced Eaton. Roger Brook Taney became attorney general. For secretary of the Navy, Jackson selected the former New Hampshire governor Levi Woodbury. As Jackson wrote to General John Coffee (who had served under Jackson during the War of 1812), "all my old friends must be abandoned by me."[19] Included in that list of friends was Branch. A few weeks after the resignations became effective, Jackson again wrote to Coffee, "The conduct of Branch has been dishonourable [*sic*] in the extreme, he ran away, or at least left me, in apparent friendship, the very day before the Publication reached here of his letter, shook me by the

3.1. Secretary of the Navy John Branch (1829–31) by Ulysses Dow Tenney. Navy Art Collection, Naval History and Heritage Command.

hand in friendship. . . . I hate a hypocrite."[20] Unlike the first cabinet, none of the replacements, except for Livingston, were sitting senators. This suggests that Jackson wanted a cabinet unaffiliated with the Washington social life, one that might show impartiality to Eaton and his wife should they return.

A month after the election of the new Congress, Jackson delivered the president's constitutionally mandated annual message on the state of the union. Jackson was cognizant of peace with European powers and the benefit that the Atlantic Ocean provided in a natural national defense. "We have nothing to apprehend from attempts at conquest," he wrote.[21] Jackson understood the reality of "attempts at conquest" and potential threats to America's sovereignty by European powers.[22] He also understood that the United States had no means of conquering territories overseas. Marines remained bound to the ships on which they served and only intermittently in the country's short history had landed ashore. As a member of the US Military Philosophical Society after the First Barbary War,[23] Jackson was familiar with Eaton's expedition in North Africa that had included a few Marines among the far more numerous Arab and Greek mercenary forces intended to overthrow the bashaw of Tripoli. Aside from a few incursions since the American Revolution, the size of the corps did not support conquering foreign territory and holding it. The Army was larger but focused on the westward expansion of the nation. In addition, the nation had no transport ships to carry sufficient troops to any distant shore. The only countries with smaller populations that the Army could theoretically conquer were in the Western Hemisphere. Mexico was the obvious choice, and less than a decade after Jackson left office, the country would mount one of the largest amphibious operations in history prior to Gallipoli during the first World War. In the case of Britain—the greatest naval power at the time—Jackson wrote of "renewing our endeavors to provide against the recurrence of causes of irritation."[24]

Jackson fully understood the need for a Navy that was adequate to protect the coast, always afloat rather than in port, and sufficiently resourced to build it up quickly in case of war. He stated that it was not the Army that would protect the country but the Navy, which furnished "the power by which all such aggressions may be prevented or repelled."[25] Jackson had clearly stated his practical vision for the Navy—to direct more funds to preserve warships already built and provide materials for future ships.

Despite the cabinet crisis in which Jackson demanded the resignations of several officials, his first term was largely successful. The Department of the Navy's finances were in a better state than in 1829.[26] Ships benefited from additional maintenance; squadrons conducted diplomatic missions and guarded against threats to American commerce. Jackson won the presidential election of 1832 with 54 percent of the popular vote and 219 electoral votes against Henry Clay's 49. Despite this, Jackson's party lost control of the Senate, even as he faced a constitutional crisis in which he would call upon the Navy.

Jackson entered office with goodwill, grand expectations from the citizenry, and dedicated support. He had a clear sense of governance that was imbued with a sense of justice (his own), honor, and fiscal accountability and responsibility. All those factors played important roles in how he managed his Navy. But Jackson chose his domestic political battles poorly whether it was the Eaton Affair or failing to work with Congress on the Second Bank of the United States. These personal shortcomings filtered down to the Navy in a government that was unable to devote more energy to the sea service. However, despite the often-chaotic machinations in the Capitol or his personal failings, Jackson managed to adapt and develop the Navy, professionalizing its ranks and preparing its ships in the eventuality of a conflict.

Secretaries of the Navy

One Jacksonian-era naval historian writes of the president's three secretaries of the Navy: "Each was selected for political reasons, and each played an unimaginative and bureaucratic role as head of the department."[27] This is arguable. They were politicians but so have been most Navy secretaries, a tradition that carries through into the twenty-first century. All three had executive experience that would be useful as the head of the department. As senators, they would have been familiar with some policy issues and especially budgets and appropriations. In nominating his secretaries, therefore, Jackson may have been more methodical and conscious of their requisite knowledge and organizational stability than previously thought.

None of Jackson's choices for Navy secretary had military service at sea. The Navy was so small that comparatively few Americans had the opportunity to serve in it; even fewer public officials had experience with it. The first secretary of the Navy, Benjamin Stoddert, for example, had been in the tobacco trade and understood trade routes, ship logistics, costs of materials, and so on. William Jones, the fourth secretary of the Navy (1813–14), had apprenticed in a Philadelphia shipyard during the American Revolution and went to sea. Each of Jackson's three secretaries had served as governors and senators, thus bringing executive and legislative experience with them to the position.

Jackson's first choice for secretary of the Navy was John McLean, a New Jersey native who, like Jackson, had moved west to Virginia, Kentucky, and then Ohio. A former member of Congress, he served on the Ohio Supreme Court and as postmaster general under Presidents Monroe and Adams. During the election of 1828, he threw his support behind Jackson and received the offers of both secretary of the Navy and secretary of war. Declining both, he received Jackson's nomination to the US Supreme Court just two days after the inauguration and was confirmed on the following day.

John Branch served as the first of Jackson's three secretaries of the Navy. A North Carolinian, he served in that state's legislature and as its Speaker of the House (1815–17) before serving a three-year term as governor. Branch then served in the Senate until assuming his office in the administration on March 9, 1829. Branch was a planter and lawyer with neither naval experience nor merchant experience with the shipping industry. Arguably, Jackson's choice for secretary of the Navy was uninspired. Branch was a political appointee and someone Jackson could trust. As a North Carolinian, he was from a state that lacked the maritime tradition of other colonies or states.

Within days of assuming the office, he and Jackson would be called upon to deploy ships. Despite Branch's inexperience, he recognized the need for changes. First, his clerks were holdovers from the previous administration. Second, he shared Jackson's intent for fiscal reform. He "fully believes that great abuses exist in the disbursement of public money under the rules or customs," recalled Amos Kendall from a meeting on March 27, 1829.[28] Branch referred to them as "derangements in finances."[29]

Despite being elected as an anti-elitist populist representing a democratic wave, Jackson relied only on long-time politicians for the top administration positions. More specifically, most were senators. It is possible that these were the individuals whom he knew and trusted since he served in that chamber. It is also possible that since their home-state legislators elected all senators, Jackson was attempting to shore up political support. None were businessmen who might have dealt with merchant ships conducting regional or global commerce.

Jackson's second secretary of the Navy, Levi Woodbury (fig. 3.2), was sworn in May 23, 1831. The New Hampshire native had served as the state's Speaker of the House, governor, and then in the US Senate before serving as secretary of the Navy until June 30, 1834, when he became secretary of the Treasury. With only a staff of ten civilians including the secretary and clerks, officers of naval stations were largely responsible for daily management of the Navy. Navy Secretary Woodbury, however, recognized the structural role of others who were employed by the Navy but held no military rank. This class included naval agents, storekeepers, builders, schoolmasters, engineers, live oak superintendents, and others. The Navy hired all these personnel on an ad hoc basis. Woodbury suggested that they instead be subject to congressional regulation and compensation. Jackson's third secretary of the Navy was Mahlon Dickerson of New Jersey, where, like his two predecessors, he had served as both governor and US senator.

The Jackson secretaries were knowledgeable, had some vision, and were amenable to the suggestions of their officers. Each made important progress unique to their tenure. Branch, for example, expected Amos Kendall to restore fiscal accountability. Woodbury used the budget to advance needed fleet repairs. His effort to create a civil service recognized challenges and

3.2. Secretary of the Navy Levi Woodbury (1831–34) by Eliphalet Andrews. Woodbury served as both a governor and senator prior to joining the Jackson administration. Navy Art Collection, Naval History and Heritage Command.

anticipated future needs. From a geostrategic perspective, he recommended the construction of two supply ships for the Pacific Station, thus reducing freight and storage costs and recognizing the permanency of the squadron. Moreover, the secretaries helped professionalize the officer corps.

Dickerson (fig. 3.3) inherited a fleet that was ready for battle, at least against like-sized powers, and he saw the advantage of steam warships. In his first annual report to Congress, six months after confirmation, he wrote, "It can hardly be doubted that the power of steam is soon to produce as great a revolution in the defense of rivers, bays, coasts, and harbors, as it has already done in the commerce, intercourse, and business."[30] He recognized that the ability to have steam warships as a means of defense would diminish costly coastal fortifications. If Dickerson was wrong, it was in his prediction that, because steam warships were too cumbersome to transport overseas for offensive operations, they would "diminish the frequency" of war. Although steamships capable of transoceanic voyage were not possible at that time, the capability might be possible in the future. In his 1835 annual report, Dickerson also recognized geostrategic threats such as those posed by Mexico and several South American nations and recommended increasing the Navy by two frigates, three sloops of war, and four steam vessels. By 1836, he thought the Navy could have twenty-five ships at sea. He later recognized the need for a new foundry to cast iron cannons as well as a national observatory—fundamental for navigation. Certainly, each secretary had limitations due to the uncertainty of the sometimes volatile president or their lack of naval experience, but they did each contribute to the Navy's evolution during the Jackson administration.

3.3. Secretary of the Navy Mahlon Dickerson (1834–38) by Arthur Conrad. Dickerson recognized the opportunities for steam warships. Navy Art Collection, Naval History and Heritage Command.

The Board of Navy Commissioners

The Department of the Navy in the early republic had very few full-time employees. The secretary of the Navy remained responsible for personnel, operations, and some administration. To assist the secretary, a Board of Navy Commissioners had been established at the end of the War of 1812. The board was comprised of three senior officers nominated by the president and confirmed by the Senate. Their duties included "procurement of naval stores and materials; construction, armament, equipment, repair and preservation of naval vessels; establishment of regulations; preparation of estimates of expenditures; and supervision of Navy yards, naval stations, and Navy agents."[31] A president governed the board.[32]

For most of the board's existence (it was disestablished in 1842), the president was Commodore John Rodgers (fig. 3.4), who presided from 1815–24 and again from 1827–37. A veteran of the Quasi-War with France, the Barbary Wars, and the War of 1812, Rodgers and Jackson were both resident members of the Columbian Institute for the Promotion of Arts and Sciences.[33] Founded by naval surgeon Dr. Edward Cutbush in 1816, it evolved into the US Botanic Gardens at the foot of the Capitol by 1838.

The board faced a challenge to its charter early in Jackson's first term. A Senate resolution sought to modify the Board of Navy Commissioners to make the secretary of the Navy the presiding officer of the board and limit the tenures of the members.[34] This would have been a radical change in governance had it succeeded. The Committee on Naval Affairs considered the

3.4. Commodore John Rodgers. No one served longer or had more of an impact on the Board of Navy Commissioners than Commodore John Rodgers. Courtesy of the US Naval Academy Museum.

resolution as well as a report issued by Senator Nathaniel Silsbee of Massachusetts. Silsbee was a rarity on the committee—or the Senate as a whole—as he had gone to sea, rising to ship's captain, ship owner, and merchant from Salem. He had been a two-term House member before being elected to the Senate in 1826 to fill a vacancy as a Federalist (later Whig). In addition to his sea service, he was the son-in-law of Benjamin Crowninshield, who had been the secretary of the Navy from 1815 to 1818 and under whom the Board of Navy Commissioners was created.

The committee did not perceive it necessary to make the proposed changes. Central to the committee's argument was that the resolution failed to specify whether the secretary would have superintendence over proceedings of the board or whether he would simply cast a vote. Regarding limiting the tenure of the officers, the committee recognized that officers of the board needed to understand a variety of subjects—construction, tactics, budgets, personnel, contracts, and so on. Rotating officers among the Navy's senior officers infused fresh perspectives, yet, the board still recognized the need for experience.[35]

The secretary waited until November 1829 to comment. This was unusual but could have been caused by one or more of three factors. First, other issues of higher importance might have distracted the administration. Second, the Navy itself or its supporters might have quietly made their objections known. Third, Jackson and his supporters in the Senate might have preferred not to expend political capital on this subject.

Branch's letter on November 13 was not to the Senate committee but rather to the Board of Navy Commissioners itself to provide guidance. He wrote that the Navy organization was "susceptible of improvement, particularly in its fiscal branch, its forms of administration, and the distribution of its duties."[36] He may have been in line with Jackson's own thoughts when he proposed "a radically different scheme of department organization."[37]

Branch also invoked Jackson's position when he advised the commissioners that he would have to submit any ideas to the president first. Such a statement empowered the commissioners. This was not unprecedented. Secretaries of the Navy during and after the War of 1812 often consulted senior officers on issues such as force structure, but those had all been Federalist administrations. Branch's outreach, therefore, suggests that the administration would at least remain consistent with how it consulted with the officers. But such power was not unlimited. In one case, a senior naval officer was admonished. Branch wrote to Jackson: "I herewith transmit you a letter I have just received from Commo[dore] Chauncey. He seems to be prematurely anticipating views of the Executive."[38]

Taking advantage of the interest in their work, or perhaps quickly responding to ensure its ideas and not others were first proposed to the secretary, the commissioners provided a comprehensive reply only ten days later.[39] The commissioners, like the secretary, appear to have recognized their own limitations as well. The duties of the Department of the Navy, they argued, were various and complicated. They stated that no one individual, however gifted, would be competent even to manage their general superintendence.[40]

Rodgers suggested a revision to the department's organization based on three clear categories. Administrative functions included the traditional duties of the secretary—the execution of regulations, orders for the fleet, and convening courts-martial. Ministerial functions focused on provisioning the fleet and ensuring an appropriate infrastructure at Navy yards.

Financial duties required a particular expertise in accounts and laws affecting expenditures. Since the position was created in 1798, the secretary of the Navy had responsibilities for administrative and ministerial duties. Financial duties, however, fell to an accountant for the Navy who was under the charge of the secretary of the treasury. Rodgers recognized inherent financial inefficiencies in the Navy. He noted, for example, that it was a regular occurrence for ships costing hundreds of thousands of dollars to require expenditures of up to $40,000 to correct errors in the structure and internal arrangements of a ship. There were errors arising solely from the absence of professional knowledge. Regarding improvement to the organization, he was entirely pragmatic in his analysis. He believed the duties of the secretary of the Navy might improve through the addition of books in his office on a variety of topics that he might consult. Rodgers meant not only books on history and geography but also books on the state and condition of the Navy so that he could answer any call from the president, or from Congress, without delay.

He suggested instead that the board develop specialties so that each officer held defined fields of expertise, a notion previously not considered in the Navy. One member would be responsible for building, repairing, and equipping warships. A second would supervise the construction of docks, arsenals, storehouses, and other shore-based facilities. The third member would oversee the provisioning of food, supplies, and clothing for the Navy. Clearly the proposal had an impact on Branch.

On January 21, 1830, Branch transmitted a plan to the chair of the Committee on Naval Affairs of the House for the reorganization of the Board of Naval Commissioners (Navy and naval were used interchangeably when referring to the board).[41] The draft legislation echoed Rodgers's recommendations about dividing responsibilities of the members: one to build, equip, arm, and repair vessels of war; a second connected with the establishment and construction of facilities on shore; and a third for naval supplies. The impact of this legislation should not be underestimated—this was the first time anyone had envisioned such a division. It set the Navy down the path a decade later to abolish the Board of Navy Commissioners and establish the bureau system that was in place from 1842 through 1966, eventually giving way to the systems commands that still administer the Navy in the twenty-first century.[42]

There was logic to having the accountant of the Navy not actually serve under the Department of the Navy and the secretary. Undue influence could occur from the secretary in cases of financial discrepancies. Instead, the law required that the individual serve in the Treasury Department as the fourth auditor. As noted earlier, Rodgers argued the necessity for an expert who understood finances and federal laws governing the budget. For this position, Jackson selected Amos Kendall (fig. 3.5), who had no experience in the topic.

Kendall's pro-Jackson newspaper, the *Argus of Western America*, helped carry the state of Kentucky for Jackson in 1828. In his articles, Kendall and others defended Jackson from adultery charges. The editor came to Washington seeking a position so he could help Jackson reform the government.[43]

When Kendall moved to Washington after the election, he sought a clerkship or auditorship. In a private interview with Jackson on February 14, the president assured him of a position as the head of a department. Kendall understood that to mean an auditorship, which would pay $3,000 per year.[44]

3.5. Amos Kendall, fourth auditor of the Treasury Department, a Jackson loyalist and member of the "Kitchen Cabinet." Library of Congress, Prints and Photographs Division.

His nomination was held back several weeks along with other controversial appointees to "avoid the hostility that would have arisen had he sent them immediately to the Senate, then in special session."[45] Half of the federal positions at the time were in the Treasury Department; therefore, it made sense that Kendall would find an assignment there. He received his commission as fourth auditor overseeing Navy accounts on March 21and began work two days later. In his autobiography, Kendall later admitted that he was totally ignorant of the process by which the Navy settled accounts. Nevertheless, Kendall possessed a sharp, inquisitive mind that quickly adapted to the accounting system. His efforts made an immediate impact on the Navy.

The decision to select Kendall for the position is curious given his lack of experience, but Jackson had ushered in a new era of government patronage. Party loyalists were rewarded for their work in the election and Kendall was in the top tier. Jackson apparently had faith in Kendall, and perhaps that is the reason he selected him to oversee the accounts of the Navy, which had had discrepancies in the previous administration. Correcting those discrepancies was a priority for the new president, particularly since the last fourth auditor, Tobias Watkins, would find himself charged with fraud in April.

Two weeks after assuming his duties, Kendall referred to his work and Watkins in a letter to his wife: "I have been so busily employed for some days in ferreting out some of the villainous transactions of my predecessor and others formerly in office here, that I have had little time for anything else."[46] Kendall later recalled: "A few individuals of doubtful integrity or dissolute habits were removed from office, and their places supplied with political friends. It was soon rumoured [*sic*] that frauds and corruptions had been detected in the late incumbents, verifying the worst suspicions of those who had opposed the last administration."[47]

Watkins was fined $3,000 and would serve nine months in jail for failing to pay. By the end of Kendall's first year in the position, he estimated he had saved $800,000 in the Navy Department, a sum representing two-thirds of the total savings of the government.[48] In November 1829, Secretary Branch asked Kendall to assess the causes of the financial mismanagement and the current laws and regulations to determine if they needed modification. Kendall replied: "Money is the sinews of power and the source of corruption. English liberty has been considered safe only so long as the power of granting supplies to the King resides in the Representatives of the people. Our institutions have gone further. Here the Representatives of the people not only grant supplies, but prescribe the objects to which they shall be applied, and the manner in which the accounts shall be kept."[49]

The financial accounting system was inefficient. No one, Kendall argued, could provide accurate information due to the poorly managed records. Nevertheless, Kendall did offer possibilities for reform. First, he recommended making the pay between Navy and Army officers equal (at the time, Navy officers were paid less than their Army counterparts). Second, he recommended specific line-item accounting for all congressional appropriations.

Kendall proved his worth to Jackson not only regarding Navy reform but throughout the two administrations as one of his closest and most loyal advisers. According to a biographer, Jackson trusted and expected only a few individuals to report to him directly—one of those was Kendall.[50] The efficiency with which he carried out his work may have led to the administration's effort to review the position and how the Navy administered its own finances. By late 1833, the secretary of the Navy suggested in his annual report to Congress that the duties of the fourth auditor be transferred to the Department of the Navy, given its close alignment. Although it was not adopted at that time, it set the stage for the later creation of the position of Navy comptroller.

Congress

Though Jackson dominated the era, he could not have done so as effectively without majorities in Congress—at least for most of his administration. If it did not share his interests, at least it did not oppose him regarding the Navy. However, in contrast to the House, the Senate was more likely to assert itself as a legislative body against the executive branch—as was the case with the reorganization of the Board of Navy Commissioners. Although Jackson was denied the presidency in the election of 1824, like-minded supporters flooded the new Congress, and their number continued to grow when he was finally inaugurated four years later.

The early republic's Congress consisted largely of businessmen and farmers who could not afford more than a few years away in Washington. In addition, congressional sessions were short—just a few months out of every year—as opposed to the nearly year-round sessions of the modern era. Members therefore gained little experience on issues, much less on the rules that governed that body. Nevertheless, members were familiar with the local militia system that had been in place since the revolution. Consequently, most members primarily focused on the Army, which offered the immediacy of protection as the American population advanced toward the Mississippi River and south through Florida. The size of the country's standing Army was only 6,000 during Jackson's first administration. In the absence of war, its greatest challenge was desertion.[51] The Army experienced more than 8,000 desertions between 1823 and 1831. In 1831 alone, 1,450 soldiers deserted.[52] Thus, Jackson and Congress understood that the Army had more immediate needs than the Navy. In 1830, for example, a report to Congress suggested that a force of 12,500 was required for a permanent peacetime Army.

Congressional activity on naval issues was minimal during Jackson's first term. A review of the *Register of Debates in Congress* for 1829–30 indicates that the Senate, as a whole, did not discuss any naval issues. The House debated only a few issues, such as an exploring expedition. Like the previously mentioned attempt to modify the Board of Navy Commissioners, discussions appeared only in the committee process and never culminated in a vote on the floor of the House or Senate.

Congressional committees also inhibited understanding the Navy. Committees were the gatekeepers who reviewed every legislative proposal and gave it their approval or disapproval. Although the nation had established a Navy in 1798, a Navy that had acquitted itself well during the Quasi-War, Barbary War, and War of 1812, Congress did not establish committees on naval affairs until 1816 in the Senate and 1822 in the House. Key committees could not establish expertise in naval affairs given their comparatively short tenures in office. Chairs of the Naval Affairs Committee had similarly abbreviated periods of service. Prior to 1846, chairmen of the Senate Naval Affairs Committee served on average only 6 months on the committee prior to assuming the chairmanship; in the House, it was 1.3 years.[53] This compares to 8.9 years and 6.9 years in the Senate and House respectively between 1875 and 1945.[54] In the twenty-first century, members might be in the Senate for decades before becoming chair and then another decade in committee leadership. Once they had become chair in the early republic, however, a member would only spend a little over 2 years in the position. Despite the lack of political experience, members knew to listen to constituents. One such example was a South Seas Exploring Expedition.

While Jackson was on his journey to his 1829 inauguration, the Senate continued considering sending an expedition to the Pacific. The proposal had been the brainchild of Ohio publisher and explorer Jeremiah Reynolds.[55] The purpose, as it evolved, was to survey the northwest coast of the US territories and the Pacific. Such an endeavor would support the American whaling fleet with navigational information and eventually expand trade with the Orient. On January 22, 1829, Senator Robert Young Hayne (R-SC) wrote to Adams's secretary of the Navy Samuel Southard requesting additional information.[56] Hayne, chairman of the Senate Committee on Naval Affairs, had just been reelected after serving a term as a Jacksonian. Like many cabinet members, Hayne had served in the War of 1812 and held a variety of elected positions including Speaker of the South Carolina House of Representatives and state attorney general. Like the cabinet, he had no experience with the Navy. His tenure in the Senate would be abbreviated in 1832 when he became governor.

On behalf of the committee, Hayne asked for clarification on several issues regarding the proposed expedition.[57] First was the issue of expenses already incurred preparing for the expedition as well as estimates for future expenditures and personnel involved. Second, the committee worried about the extent of the expedition's goals and if it might take several expeditions. Southard referred to a resolution of the House of Representatives from May 21, 1828, stating that the expedition was to "examine the coasts, islands, harbors, and reefs in those seas and ascertain the true situation."[58]

The country's nautical charts were extremely inaccurate, endangering the safe travel of both commercial and naval ships. Southard reported that whalers believed there were at least two hundred island reefs and shoals that were not noted on any American chart. The Navy proposed sending only one of its smallest ships—the sloop *Peacock*—for this assignment. Supernumeraries would include an astronomer, naturalists, draftsmen, surveyors, and someone to assess the state of commerce in the Pacific.

The Navy intended to enlist experienced deckhands from the whaling centers of Nantucket and New Bedford and had already sent out recruiters. The Navy made repairs to the *Peacock*, contacted scientists, and consulted Americans with experience in the Far East. In short, the Navy had laid the essential groundwork, should the expedition receive approval. The House of Representatives had already authorized $50,000 for the expedition, but the *Peacock* alone, the Navy knew, could not conduct the expedition in its entirety because of the extent of territory involved in this mission.

Less than two weeks before Jackson's inauguration, the House and Senate differed on the expedition. Hayne reported to the committee that the House had expected the Senate to concur with the House's resolution without due consideration. Hayne recommended delaying action on the resolution in favor of a future, much larger, and more expansive project. Hayne appears to have argued that the cost of rebuilding, repairing, and fitting out the expedition, in addition to expense of maintaining it for three years, was "very considerable."[59]

More troubling to Hayne was the fact that the Navy had progressed so far with its preparations without Senate concurrence on spending. Constitutionally, the Navy was supposed to react to congressional guidance, not initiate it. He found further investigation was necessary. The committee could not support the proposal and suggested instead the country had too many "unsettled and unexplored regions at home."[60] The country had to push westward and take the lands up to the Pacific Coast before such a maritime endeavor could begin. Hayne and the committee recognized the opportunities an expedition offered, such as the possibility of discovering new islands that might serve as the foundation for an American colony. The report also referred to Washington's warning in his farewell address that the country should avoid entangling alliances. He cautioned against a "spirit of adventure, without an abandonment of the fundamental principles of policy, in a departure from those wise and prudent maxims which have hitherto restrained us from forming unnecessary connections abroad."[61] Asserting senatorial privilege, Hayne proposed a resolution that Jackson provide a detailed statement on the preparations for an expedition.[62] However, Jackson had other concerns with the Navy and the country. The expedition would have to wait another eight years, when a six-ship force finally weighed anchor for a four-year cruise.

The midterm election of 1830 had mixed results for Jackson. In the Senate, Jacksonians continued to hold a two-seat majority over the anti-Jacksonians. In the House, however, Jacksonians lost ten seats. Like the Congress of 1829–30, the new Congress spent little time debating naval issues. Instead, it focused on domestic and fiscal issues, such as the post office and tariffs. When Congress did raise naval issues, it tended to focus on pay. Jacksonians supported increasing the pay of masters commandant, but not of lieutenants, thereby further distinguishing between the two ranks.[63] This was the first indication of a clear separation of those ranks, particularly since master commandant would evolve into the rank of commander by 1838. As part of the administration's expectation for accountability in the Navy, regulations were proposed that governed prices for necessary articles on board ships, thereby providing consistency and professionalizing the role of purser.[64]

The Budget

One of the clearest pieces of evidence that show a presidential administration's priorities is the annual budget. This was no less true of Jackson. He inherited a naval budget in 1829 of $3 million; his last budget of 1837 was more than $5 million (fig. 3.6). A fiscally conservative president less inclined to support the Navy would not have increased its budget by 270 percent in only eight years. Yet, the spending categories within the budget were not consistently increased. They were subject to changing circumstances, such as the Second Seminole War late in Jackson's administration, the construction of new ships, particularly the costly ship of the line USS *Pennsylvania*, and improving pay for active-duty personnel.

Jackson entered his first term with the intent to hold the Navy Department fiscally accountable and adjust for perceived mismanagement. The full effect of a president's budget is felt approximately eighteen months after taking office, as the first budget is inherited from the predecessor. Jackson, for example, modestly increased the budget in 1830. However, after receiving reports from Fourth Auditor Kendall and having a full understanding of naval spending, Jackson responded at the first opportunity to impose austerity measures. From 1830 to 1831, the Navy budget decreased 25 percent. This was achieved by slashing the budget for repairs and improvements at the Navy yards by nearly 50 percent (although it was restored the following year) and for provisions by more than 60 percent and by zeroing out ordnance procurements for one year. These cuts were accompanied by moderate increases in some categories, such as naval pay.

By December 1833, Jackson felt confident enough with federal spending that he could respond to an earlier Senate proposal to equalize the pay of officers of the Army and Navy. One method of achieving this was to decrease the pay of senior officers. The senior Army officer held the rank of major general. While the Navy would not have the rank of admiral until the Civil War, its senior captain while afloat would receive the same pay as the major general, for

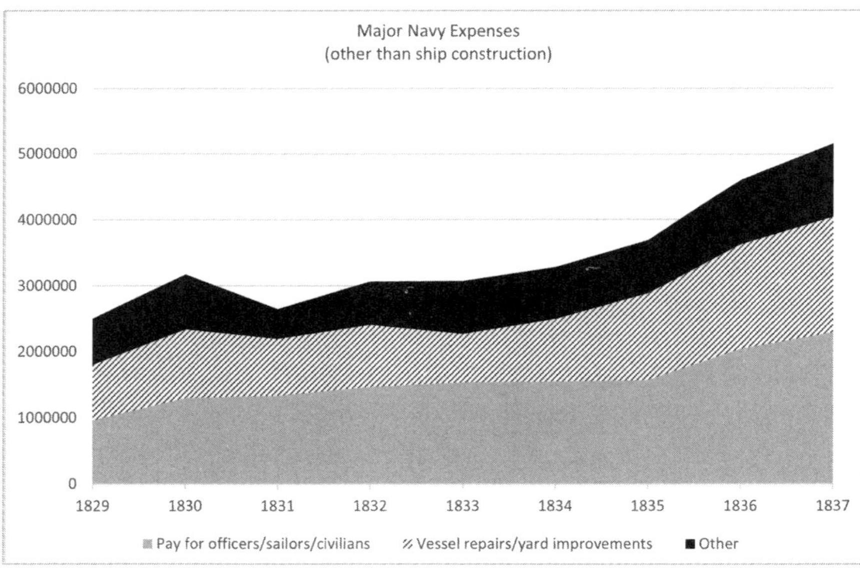

3.6. Navy budgets, 1829–37. Annual Reports of the Secretary of the Navy, *American State Papers*.

example. Naval pay became a major component of his professionalization of the sea service, with an increase of 270 percent from 1829 to 1837.

By 1835, because of the boom in revenues due to public land sales, partly due to reducing their price, the president did what no other president in US history achieved—he balanced the budget and eliminated the federal debt. This enabled Jackson to then focus on new budget priorities, such as those for the Navy. In addition to naval pay, Jackson reinvested in improvements at Navy yards, spending nearly $1.2 million in 1836 and 1837 alone. He also invested in the repair and construction of ships as well as in a return to pre-1830 levels for ordnance. Admittedly, ordnance would be a higher budget item in 1836 and 1837 because of the Second Seminole War. The benefits of these investments and adjustment of priorities are discussed in the following chapters.

Jackson's governance was guided by a clear sense of where America should go—to the sea, to engage in global commerce to spur the economy. He understood that the Navy was required not only to protect US merchant ships but to conduct reprisals if necessary. It is clear from inauguration speeches and his annual addresses to Congress that he believed the Navy must advance technologically and increase in size. To do so, he first had to address budget issues, which he accomplished.

Due to his personal sense of justice, Jackson held the Navy accountable for transgressions by sailors and officers. He was more involved in court-martial reviews (discussed in a later chapter) than any president before him or after him through the presidency of James Buchanan. But Jackson's personal emotions could cloud a more logical and fair approach to an issue.

Jackson did much during his two terms, especially for and with the Navy. Nevertheless, some of his efforts were marred or stunted by self-inflicted political wounds based on a poor temperament. These included his reactions to the Eaton Affair, nullification, and the Bank of the United States. It would be simply speculative to suggest what he might have accomplished with the Navy without these distractions to his governance.

One historian contends that "there is no evidence that Jackson and his supporters devoted a substantial amount of time, reflection, or discussion to naval policy."[65] At face value, this statement is correct. There are no recorded discussions of policy, nor are there substantive letters from Jackson to his secretaries of the Navy on issues. But despite the paucity of direct correspondence, it should not be taken there were no discussions or no thought. Public statements like annual addresses and reports point to the contrary, and actions speak louder than words. Subsequent chapters on platforms, operations, and personnel will show what was achieved.

4

Force Structure and Modernization

Jackson oversaw the restructuring and modernization of the Navy during his presidency. His predecessor, John Quincy Adams, had surged ship construction and employed squadrons worldwide. Rather than steer a different course, Jackson continued to expand the Navy and its squadrons. To do so, he first invested in the maintenance and repair of existing warships. He then continued the Adams-initiated construction of the first dry docks and other infrastructure. Jackson then authorized the construction and purchase of sloops and schooners, which enabled him to continue high deployment rates. Despite the utility of these ships, he permitted construction for prestige purposes of the second largest ship of the line in the world. Finally, the Jackson administration endorsed the advancement of steam technology for warships to modernize the Navy.

Several of Jackson's predecessors—John Adams, Thomas Jefferson, James Monroe, and John Quincy Adams—had served as ambassadors to European courts. Consequently, they had at least some firsthand understanding of merchant and Navy ships. More than 3,500 nautical miles of ocean separated the United States from Great Britain. These individuals had stood on the decks of warships. They had seen American sailors who climbed the rigging and who ate food for sustenance rather than taste. They spoke to the officers who commanded the men. When they arrived in European ports, they saw before them the might of maritime powers and the navies built to defend merchant fleets and deploy forces. It is not surprising, therefore, that the establishment, construction, and use of a Navy for the United States would be significant during their presidential administrations.

Unlike those predecessors, although he was familiar with the Navy through the Battle of New Orleans, Jackson had no such similar grand experience since he never visited a European port. This is not to say he did not understand the Navy. But his experiences were different from his predecessors as they were far more parochial. Nevertheless, he may have been shaped by his exposure to navies beginning with his youth.

The first may have originated with the British siege of Charleston, South Carolina, in 1780, when the British Navy surrounded American land and naval forces, while American ships made their way upriver to prevent their capture.[1] The British force included three ships of the line; four frigates, sloops and galleys; and approximately ninety transports.

He was only thirteen when his relatives asked Jackson's mother to permit him "not to fight, just to help out" in operations against British forces.[2] In the subsequent Battle of Hanging Rock, an elder Jackson "declared himself unable to give an account of the battle. Not only had he not participated in the fighting, he had not even seen it."[3] The subsequent capitulation of the American force at Charleston would be the largest of the revolution and the third greatest in American history. American soldiers were held on prison ships in the harbor.[4] In 1781, Elizabeth Jackson succumbed to the same cholera outbreak that affected those in her care. She left her son an orphan. At his young age, Jackson would not likely have distinguished between the classes of ships present in Charleston harbor. Instead, the Royal Navy represented an unstoppable force. It prevented American ships from escaping through an effective blockade, and it brought supplies and soldiers ashore that had a direct effect on his family.

A second exposure took place in the waning days of the War of 1812, as General Jackson again fought the British. Jackson awaited British admiral Alexander Cochrane, who had failed to take Fort McHenry and Baltimore, Maryland. Cochrane was unable to convey his deep-draft ships through the shallow waters of Lake Borgne, defended by Lieutenant Thomas Catesby Jones's five Jefferson-era gunboats.[5] Cochrane deployed forty-two longboats with carronades, which eventually overwhelmed Jones's force. Lake Borgne was a tactical victory for the British but a strategic loss.[6] Jones's action provided Jackson time to assemble his own forces and prepare defenses against General Sir Edward Pakenham's Army of 14,000 men. Lake Borgne and New Orleans represented to Jackson the value of the Army and Navy working jointly in combat operations, a view that would reemerge during the Jackson-initiated Second Seminole War.

The third influence on Jackson's maritime view may have been his journey to his inauguration. Most of his trip from Tennessee to Washington was on America's extensive river system aboard steamships. This trip reinforced upon Jackson the vital role riverine and other waterborne commerce already played in the country's economy. International commerce was vulnerable to pirates, and competing navies also threatened the lifeblood of America's economy. Consequently, a standing Navy during peacetime was a necessity. In addition, he observed—not for the first time—the value of the nascent technology of steam power for ships. Jackson would have recognized that they were thus far insufficient for transoceanic travel but in the near term could be useful for coastal or harbor defense.

The Evolving Navalist

One historian's precedent-setting work defined the navalists and antinavalists.[7] A major distinguishing factor between the two philosophies was the number of ships authorized or constructed. However, the authorization and construction of Navy ships were irregular during the early republic from the founding of the Navy under President John Adams through the end of Jackson's presidency. An analysis of the shipbuilding program during that period reveals that presidential administrations—regardless of party—were mostly navalists especially John Adams and John Quincy Adams. A closer assessment suggests that Jackson should be included among the navalists.

The US Navy built a total of eighty-six warships from 1798 to 1837.[8] Most shipbuilding occurred during armed conflicts. During this forty-year period, there were four spikes in ship construction that reached their apex with a ten-ship build. The first spike was a result of the Naval Act of 1794 and the creation of the Department of the Navy under John Adams. The bill was passed under threat of Algerine piracy and authorized the construction of the first six frigates designed by Joshua Humphreys, but it prepared the country for a maritime conflict with France. The second spike occurred during the second war with the British. The small US Navy was still reliant, given the immediacy of military operations, upon privateers until the construction of more ships. Due to a lack of resources and the length of construction, the Navy would commission ships of the line only after the war's completion. Congress thus passed "An Act for the Gradual Increase of the Navy" in 1816.[9] This legislation authorized the construction of nine 74-gun ships of the line and twelve 44-gun frigates. Under the Constitution, funding requires authorization and appropriation. Authorizing legislation provides policy guidance and determines that Congress may fund a program. Nevertheless, Congress may authorize a program, but it remains unfunded if a separate committee refuses to appropriate the funds in a designated fiscal year. Congress did not fully fund the authorizing legislation, instead only providing sufficient funds to purchase materials and store them in shipyards for most of the ships, while fully funding a few ships of the line and frigates. Enough, however, were commissioned by 1821 to result in the third spike. The fourth spike came under the presidential term of John Quincy Adams. Adams commissioned thirteen warships—no small feat given that he was only in office for four years whereas his three immediate predecessors had each served for eight years. Adams's construction program, thus, is the largest peacetime increase of warships during this period.

A lull followed each construction spike. This is not unusual in the United States or for other nations in postwar environments. Military budgets increase exponentially during war, and, when the war is over, those expenses are no longer required during peacetime. After the Quasi-War and First Barbary War, Jefferson in 1805, for example, decommissioned several larger ships to reduce government expenditures. Again, the Navy represented a sizable portion (from 20 to 30 percent) of the federal budget, so any decrease in the Navy meant a president could be more fiscally responsible in peacetime. Only the navalists supported an expansion of the Navy in peacetime. Consequently, Monroe's administration advocated an increase to establish more distant squadrons to protect American commerce. This squadron system also enabled a faster response time to overseas threats.

The number of ships built also sheds light on the views of the presidents as well as the country's priorities. Of the ninety-one oceangoing warships in the Navy from 1798 to 1837, American shipyards built fifty-one during periods of wartime operations and forty during peacetime (fig. 4.1).

That 44 percent of America's warship construction occurred during peacetime is remarkable. This suggests not only a necessity for peacetime ships but a commitment to build and maintain them. America was, therefore, largely a maritime-oriented nation, willing to spend the money to defend its commerce. The depredations committed on US shipping by the Barbary States in the 1780s and 1790s and threats from greater powers were not lost on the American public. Washington paid tribute to the Barbary States out of the

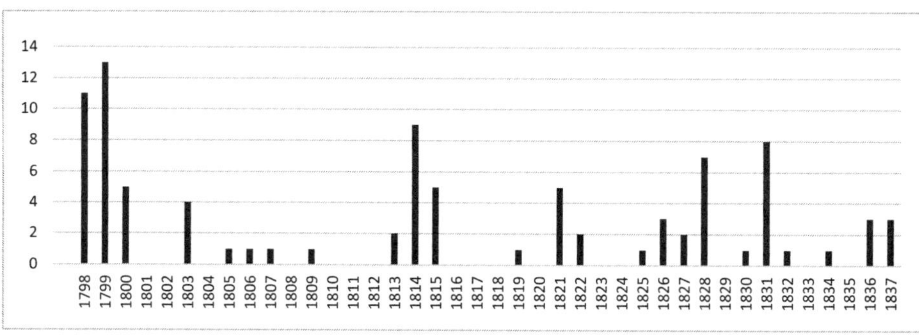

4.1. US Navy ships built, 1798–1837.

federal budget. The average American, however, was uninformed about government policies and budgets. Normally, they might have ignored it. In this case, the American public was very aware of the reality given that the local level directly felt the ransoms paid for seized crews.

Jackson's administration confirms that he was a navalist based on three quantifiable criteria of ship production. The first data point is the average number of ships built annually. From 1798 to 1837, an average of 2.3 ships were built annually. During war years the annual number of ships built was 4.6 and during the peacetime years the number was 1.4. Jackson built 2.1 ships per year, significantly higher than the overall comparable peacetime average. If one removes Jackson's eight years, the average of the remaining peacetime years' average is 1.2 ships built per year thus widening the construction gap compared to Jackson's 2.1.[10]

The second way of viewing the data is by the total number of ships built by administration. Remarkably, John Adams built twenty-nine ships during the creation of the Department of the Navy, a period that also coincided with the Quasi-War with France. Adding to this impressive achievement was that he served for only four years and only his last three years are included in this study. Madison (who had three years of war during his eight-year administration) built seventeen ships—the same as Jackson. Behind them was John Quincy Adams whose four years produced thirteen ships with the other presidents (Jefferson and Monroe) in single digits.

The third set of data is the average number built annually by each of the period administrations. Again, Adams led them all with 9.7 built on average per year. Distantly behind him was his son John Quincy Adams (3.3). Madison and Jackson each built, on average, 2.1 ships annually. Seen from each perspective, the data demonstrates that Jackson had a deep commitment to the Navy and its mission that was equal to the navalists and exceeded that of the antinavalist presidents.

Setting Priorities: Maintenance and Repair

When Jackson became president, he did not immediately embark on a shipbuilding program. Instead, he wisely chose to invest in the maintenance and repair of the existing, aging fleet. For example, four of the Navy's original six frigates—*Constitution*, *Constellation*, *Congress*, and *United States*—were still considered operational (at least fundamentally) after thirty years and three wars. Of the fleet's ten operational frigates, only three were less than a decade old when Jackson assumed office. In 1822, the average ship age was 9.4 years.

When Jackson became president in 1829, the average was 11.6 years. In the final year of his administration, it was 17.7 years. Moreover, after the War of 1812, the Navy was maintaining distant stations for longer periods of time. It was not unusual for ships to deploy for two to three years on patrol. In one case the ship of the line USS *Franklin* deployed as the flagship of the Mediterranean Squadron from 1817 to 1820. It returned to the United States for a refit only long enough to prepare it for another three-year deployment from 1821 to 1824 as the flagship of the Pacific Squadron.[11]

Jackson arrived in Washington intent on bringing fiscal responsibility to the US government. In Branch's first annual report to Congress, he articulated the administration's position that ship construction would discontinue except for in the case of an immediate emergency. In 1829 the total Navy budget was $3 million—approximately 20 percent of the entire federal budget. Of that $3 million, $430,000 was directed for "repairs of ships in ordinary and wear and tear for ships in commission" (hereafter maintenance and repair). The following year, maintenance and repair increased to $590,000, as the Navy budget rose to $3.5 million.

Jackson's austerity measures and resolution of financial discrepancies by Fourth Auditor Amos Kendall allowed the president to reduce the Navy budget to $2.6 million. Remarkably, maintenance and repair funding increased to $615,000. By the final year of his administration in 1837, the maintenance and repair budget increased to $1.2 million. Why is this important? Jackson was in fact enabling the Navy to provide a far larger force, if called upon. This would be accomplished by building new ships, while ensuring aging ships or ships that remained in ordinary—those authorized but not fully funded vessels that lay in various states in shipyards on the Eastern Seaboard—received adequate attention.

When Jackson assumed the presidency, he had at his disposal nearly three dozen commissioned ships. Many were deployed. Each year the secretary of the Navy, based on assessments from the chief naval constructor, reported to Congress on the state of commissioned ships. The assessment also included ships in ordinary. Some ships had their keel laid, while others had partially constructed hulls. Some materials remained the victims of weather, as they were stored in the open air given the limited availability of storage sheds.

The most expensive ships in terms of cost and manpower were the 74-gun ships of the line. Of the six *Delaware*-class ships of the line, all had been laid down. Due to excessive cost, two were commissioned in the 1820s, two were only commissioned in the 1860s during the Civil War, and two were never launched or commissioned. Construction was not the only significant cost. So, too, was the price of operating the ship in distant waters. In 1829, for example, the USS *Delaware* in the Mediterranean required more than 300 barrels of flour, nearly 3,000 barrels of bread, 1,200 barrels of beef and pork, more than 2,000 gallons of vinegar, 1,000 gallons of molasses, and more than 24,000 gallons of whiskey for a three-month period.[12]

In the 1829 report, Secretary Branch assessed the *Alabama* could be ready in three months, if needed. Navy constructors assessed her hull to be in good order. In Boston, the ships of the line *Virginia* and *Vermont* were both four months from completion. *Columbus* could be ready in seventy-five days, while the *Independence* required entirely new decks, planking, and coppering. In Gosport (near Norfolk) the ship of the line *New York* could be launched in three months, but the *North Carolina* had had all her oakum removed and had no estimated time for completion.

At the New York Navy Yard, the commissioned and operational ships of the line *Franklin*, *Ohio*, and *Washington* all needed extensive repairs.[13] The coppering of the *Ohio* alone required six months. Far different was the ship of the line at the Philadelphia Navy Yard. She was the 120-gun ship *Pennsylvania*, also authorized in 1816. For more than a decade, she was in pieces, awaiting funding for construction and preparation for sea. In the 1829 and 1830 annual reports, she was assessed to be ready in six months. She would have to wait until later in Jackson's administration for determination as to her fate.[14] The entire class, however, failed to evolve under constructor Samuel Humphreys. The situation with the state of American frigates was different from the large ships of the line. All eight *Potomac*-class frigates authorized in 1816 were eventually commissioned, although only one was commissioned prior to Jackson's presidency and one in 1831. Across the shipyards, the *Santee, Cumberland*, *Constitution*, *Raritan*, *Sabine*, *Savannah*, *Potomac*, *St. Lawrence*, and *Constellation* were all in varied states of construction or repair.

Assessments differ on the quality of shipbuilding, maintenance, and approaches in the Jacksonian era. One example was the frigate USS *Macedonian* in Gosport Naval Shipyard in 1829. One historian has argued that the *Macedonian* had been unserviceable for years, but she remained on the registers to obtain annual maintenance funds from the Navy.[15] According to the February 1829 assessment, the ship was in a state of decay, especially the white oak—the timber used specifically for Navy ships since the Humphreys frigates. Chapelle, however, did not consider that the *Macedonian* had only recently returned from deployment in 1828 or that the USS *Constellation* and USS *John Adams* had already been planned for refurbishment.[16] Funds were available but the docks at which the work would be performed were not available. In an unpublished manuscript, another academic researcher challenged the earlier work. Secretary of the Navy John Branch apparently opposed work on the ship to streamline the Navy itself. Reducing the number of ships meant a reduction in annual maintenance costs. Nevertheless, Secretary Woodbury added a request for more than $200,000 to rebuild, repair, and equip the *Macedonian* in 1831. The key term in that direction was "rebuild." Branch's interpretation was that rebuilding meant simply new construction, which required a supplementary budget request to Congress.[17] A few years later the *Macedonian*'s keel would find use in the construction of a new ship with the same name.

This confusion over rebuilding was not isolated to this event. Two decades later, for example, the USS *Constellation* had most of the planks and ribs replaced and retained only the keel. The ship consequently had a far different design, particularly with her new rounded stern indicative of mid-nineteenth-century British-inspired alterations, rather than early nineteenth-century standards. Historians have also failed to adequately consider the impact of patriotism behind the support of shipbuilding. In 1832, Woodbury asked for funds for more live oak to rebuild the 22-gun USS *Cyane* and for additional funds for the frigate USS *Java*. *Constitution* had taken the British-built *Cyane* and *Java* as war prizes during the War of 1812. *Macedonian* was also a Royal Navy ship captured in 1812 by *United States* under command of Stephen Decatur. *Constellation*, though better known for her service during the Quasi-War, had also fought in the War of 1812.

Thus, despite their various states of disrepair, their names alone conjured up patriotic sentiment. Proponents of the funds appealed to the patriotism of other congressional members. The War of 1812 represented the most recent conflict during which American frigates largely bested Royal Navy ships in single-ship actions. Support for older ships continued through Jackson's administration. In 1834, for example, Congress approved another special appropriation of nearly $200,000 to rebuild *Congress*—another Humphreys frigate. Despite this initial support, the ship proved too unstable and was broken up that year.

Early in the administration, Branch recommended to Congress that it should take a twofold approach: first, consolidate the ships that could be completed, and second, delay the construction of new ships. Branch used one ship in particular to make his case for delaying construction so that he could focus on maintenance. The 44-gun frigate USS *Hudson* was the only ship in her class. Originally named the *Liberator*, she had been constructed in 1826 for the Greek government for Greece's war of independence. When that government failed to pay for the ship, the Navy retained and renamed her. She was fitted out in 1828 for a cruise to Brazil Station. The estimate for fitting her out was $51,770. This included $15,000 for labor and materials; $1,800 for the boatswain's department; $15,000 for deficiencies in cables, anchors, hawsers, and other equipment; $12,400 for the gunner's department; $5,400 for the sail maker's department; and $1,900 for the cooper's department.[18] On the ship's first cruise Captain Stephen Cassin reported the ship constantly required repairs of some kind.[19] "I regret to state," he wrote to Branch, "that the ship throughout is in a very rapid state of decay, and in my opinion, cannot be kept out longer than May or June next with safety."[20] In addition, although food was rarely optimal on Navy ships, Cassin also reported that he would not be asking for the staples such as molasses, rice, cheese, and beans. "From experience," he added, "it has been found that some of these articles are of so perishable a nature, that heretofore . . . most instances unfit to be served."[21]

The *Hudson* was a troubled ship structurally. Ships with significant structural and maintenance problems also manifest their problems in the morale of their crews—an occurrence as familiar to the twenty-first-century Navy as it was two hundred years ago. Therefore, it is important to assess shipboard life during its cruise. One method of assessing morale is through courts-martial (fig. 4.2). This is both an effective quantitative and qualitative means. The reasons for the courts-martial and the text of the records illustrate the qualitative element. One can establish with the quantitative element comparative number of courts-martial throughout the fleet.

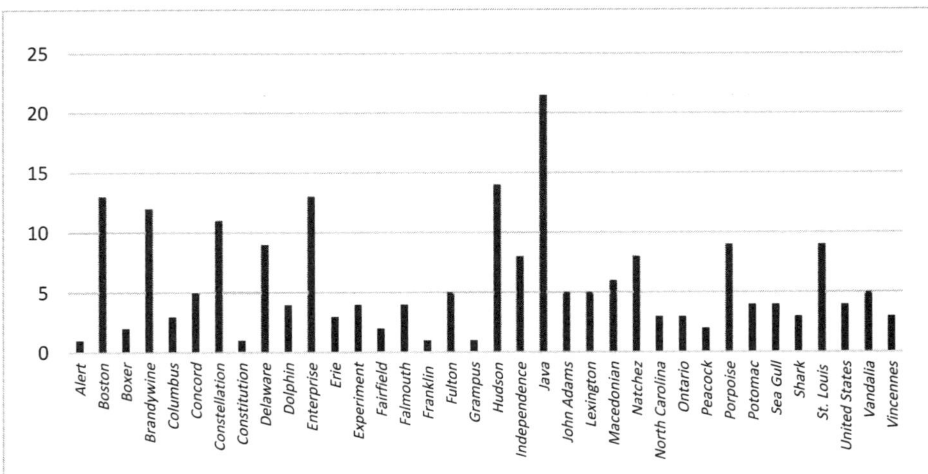

4.2. Navy courts-martial, 1829–37, by ship. Compiled by author from Records of General Courts-Martial and Courts of Inquiry of the Navy Department, 1799–1867, Navy Courts-Martial, 1825–1840, National Archives, Record Group 125, M273.

Courts-martial abounded aboard USS *Hudson*. Rank did not protect anyone, and behavior was not indicative of rank. Sailors and officers alike found themselves pulled into this vortex of personal irresponsibility, professional negligence, and internecine irresponsibility. Sometimes these courts-martial were about Jackson himself. For example, Sailing Master Lieutenant Thomas Hamersley was charged in a case of political disagreement with a British officer in Rio de Janeiro and using disrespectful language about his own commodore, John Creighton. Hamersley appealed to the secretary of the Navy arguing that the root cause was due to his political sentiments, namely that the junior officer supported Jackson in the election. When challenged by the slanderous British officer, Hamersley had simply attempted to vindicate Jackson, whose reputation he considered "a National property."[22] In his defense, Branch wrote to Jackson that there was no improper trial and that the argument with the British officer did not stem from improper motives "but was rather prompted by feelings of national pride."[23] Jackson approved Branch's recommendation to retain the lieutenant.

Given the number of courts-martial for such a small ship, it is difficult to trace the extent of charges by crewmembers against other ship's personnel. Surgeon Andrew Cooke charged Marine Second Lieutenant Francis Neville for conduct unbecoming an officer, gross falsehoods, and scandalous conduct, specifically for calling Commodore John Creighton "a damned scoundrel" and a villain.[24] Neville was found guilty and dismissed from the squadron.[25] Midshipman William Moore placed similar charges against Midshipman Ferdinand Smith for striking another midshipman. Found guilty, he was sentenced to be cashiered. Only three individuals could approve, modify, or nullify a court-martial—the commodore of the squadron, the secretary of the Navy, or the president of the United States. Creighton deemed it a harsh sentence. Jackson eventually concurred and ordered the midshipman returned to duty.[26]

Lieutenant Tom Freelon was charged with disrespect toward Commodore Creighton and sentenced to dismissal from the squadron. Again, Jackson interceded questioning the validity of the court as it served as both accuser and witness and Freelon was ordered back to duty.[27] Lieutenant Joshua Sands was tried for neglecting to enforce regulation of the ship and using menacing words against the surgeon.[28] Surgeon Henry Bassett was charged with falsehoods and conduct unbecoming an officer.[29]

The results of each court-martial quickly spread as junior officers in particular wrote to one another. Such was the case of Surgeon John Wiley writing to Lieutenant Samuel Du Pont: "You have heard doubtless of the quarrels and court-martials on board the *Hudson*, but you will be surprised to learn that Capt Creighton's particular supporter and judge advocate on all the trials has himself been dismissed from the Navy by the President for personal abuse of the chief magistrate."[30]

Unusually troubled ships did not escape the attention of officers. Surgeon John Wiley wrote of the quarrels and courts-martial aboard the *Hudson*.[31] Wiley likely referred to at least two trials, both held in the harbor of Rio de Janeiro. The first was the aforementioned trial of Marine Second Lieutenant Francis Neville. The second was the trial of Midshipman William Moore for provoking and reproachful language for calling another midshipman a liar and a "son of a bitch."[32] Moore was sentenced to be cashiered. Although Jackson reduced the sentence, Moore resigned in 1832. Wiley was correct to be concerned about the *Hudson*. While on Brazil Station, no fewer than seven courts-martial were held involving its officers. Another eight courts-martial involving its officers and crew were held when the ship returned to New York.

One cannot discount that the ship suffered from a lack of leadership. One historian suggests that Creighton had the "reputation of being the greatest martinet in the Navy" but one who held the importance of appearances and drilling.[33] While commander of the ship of the line *Washington* in the Mediterranean, Creighton struck Midshipman John Marston. Marston brought charges against Creighton, which were dismissed as "malicious, frivolous and vexatious."[34] After a similar incident with Commodore Oliver Hazard Perry striking a midshipman, fifty-one midshipmen wrote to Congress asking it to investigate.[35] Navy captains responded by demanding the readjustment of midshipmen. The Senate Naval Affairs Committee largely ignored the issue, instead suggesting that the Navy handle its own affairs of justice.

In the decade from 1828 to 1839, the Navy held 309 courts-martial. Nearly twenty of those were on the *Hudson* alone, which was the second-highest number of any ship in service during that period. The only other ship with more courts-martial was the USS *Java* with more than twenty-five. In 1832, Secretary of the Navy Levi Woodbury asked Congress for additional funds for the *Java*—a request made specifically because of its own significant problems. The *Java* had been built in haste as a 44-gun frigate during the War of 1812. When it returned from a cruise in 1831, it was made into a receiving ship in Norfolk and was unable to put to sea again. Built of inferior materials, the ship was assessed as defective in all respects.

Another source on officers also communicated about the state of the ships and their captains. Mackay lamented the fact that the *Hornet*'s government of the ship was "weak and imbecile to a degree of which you can form no conception—the premier . . . is totally incompetent and disqualified in every point of view."[36] Of the USS *Warren* in 1836, one officer wrote that it had gone to sea without a proper overhaul—the spar deck leaked, parts of the ship were rotted, and the trucks of the guns were untrustworthy.[37] While it could be argued that every sailor before or since has complained about their ship in some way, the high deployment rate of sloops immediately before and during Jackson's presidency meant significant maintenance issues that were reflected in the observations of the junior officers.

By 1836, the only two ships deemed unfit for service by the Board of Navy Commissioners were the *Hudson* and the *Java*.[38] Two key assessments can be derived by this fact. First, there was a causal relationship between significant maintenance and repair issues with crew morale. Second, Jackson's early emphasis on maintenance and repair investments—especially in the first three years of his administration—yielded positive results in the fleet. Had he not done so, it is likely that additional ships would have been deemed unfit for service by the end of his second term. Instead, Jackson left office with a Navy better prepared to conduct peacetime and potential wartime operations. Maintenance and repair of ships were only some factors in ensuring the fleet was ready for long-term deployments or at domestic bases awaiting orders. The Navy required better facilities.

Dry Docks and Navy Yards

To better prepare the Navy for long-term operations, Jackson supported the continued construction of the first dry docks and increasing the efficiency of the Navy yards. This enabled quicker and cheaper service of the ships. Based on input from the Board of Navy Commissioners in 1829, Branch found that the Navy was cutting the best timber in the country for authorized ships that were not being constructed, and thus, the timber was being exposed to the elements. According to the board, the resulting decomposition was ruinously wasteful. Branch conceded that Jackson's intervention was required. The administration had to decide.

Most of the Navy yards were nearly three decades old. Portsmouth, in New Hampshire, was built in 1800; Boston and New York were established in 1801. Norfolk had only recently been established in 1827 with a new dry dock in the early phases of construction. The Washington Navy Yard (fig. 4.3) was the oldest shore establishment, built in 1799, and commanded by Commodore Thomas Tingey from its opening until a month before Jackson's inauguration. Even the shipyards were not immune from political machinations

4.3. *City of Washington from beyond the Navy Yard*, c. 1833, by G. Cooke. Though the republic was nearly fifty years old, Washington, DC, remained a small city dominated by the navy yard, the Capitol, and the White House. Library of Congress, Prints and Photographs Division.

and debates. In 1829, there was at least one incident reported to Branch from the Philadelphia Navy Yard leveling charges against a senior civilian worker regarding "improper influence at the last election" and of "abusing General Jackson and his deceased Lady."[39] In 1834, Charlestown Navy Yard in Massachusetts became the scene of one of the most tangible outrages.

Captain Jesse Duncan Elliott was one of the Navy's "Jackson men."[40] He was appointed commandant of the Charlestown Navy Yard in 1833. In his own words, he revered Jackson and in 1834 to honor him had a figurehead of the president made for and mounted on the bow of the USS *Constitution*. Doing so in the Adams's family backyard was tantamount to treason in the eyes of his detractors, who threatened to tar and feather him.[41] The ship was nestled between the ships of the line USS *Columbia* and USS *Independence*. Bostonian Samuel Worthington Dewey (fig. 4.4) took it upon himself and rowed to the *Constitution*'s bow during a thunderstorm and sawed off the head. Elliott wrote of it that a "hireling from the darkling brook of party ruffians stirs with a dagger the ashes of a hero's grave."[42] Elliott claimed further desecration of the figurehead: "The head of the image was brought in, laid upon the table, and bacchanalian orgies were held over it!!!"[43]

In Branch's first year as secretary, he pursued cost-savings measures with the Navy yards recognizing their existence as one of the costliest functions in the Navy. He again turned to the Board of Navy Commissioners to investigate, asking if the budget could be materially diminished by a reduction of their number.[44] The board took a strategic approach to the study based on geographic threats and technologies. Based on geographical positioning and historical experiences from the revolution and War of 1812, Chesapeake Bay and Narragansett Bay (off Rhode Island) were the most vulnerable to foreign incursion in what its members called "our maritime frontier."[45] The board envisioned a spirited defense of those frontiers with both ships of the line and steam batteries. It deemed Portsmouth as the least defensible. New York, Philadelphia, and Pensacola were assessed as not accessible enough. Washington had larger ships, but they could be moved by steamboats. The board's report recommended retaining the Navy yards at Boston, Washington, and

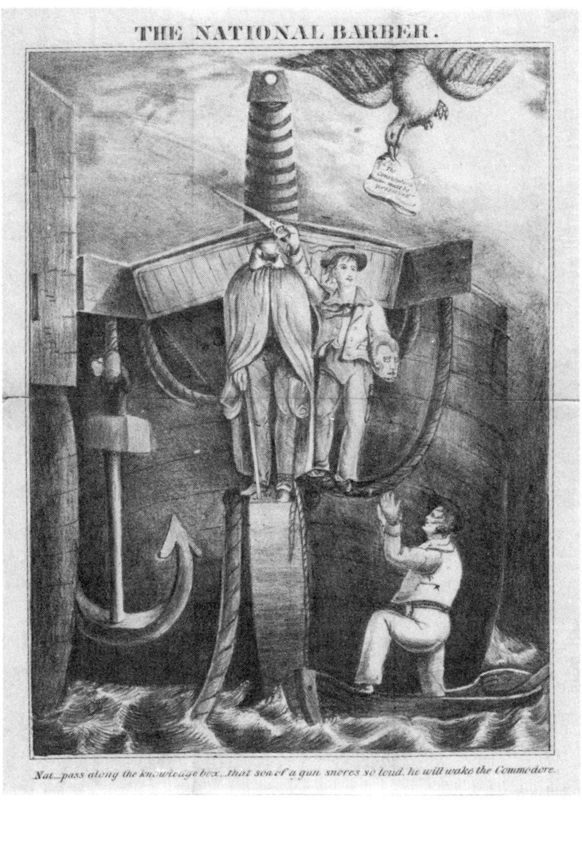

4.4. *The National Barber.* Print shows a sailor, identified elsewhere as Samuel W. Dewey, having sawn the head of Andrew Jackson off the figurehead of the USS *Constitution*, pass it to another sailor awaiting in a rowboat alongside the ship. An eagle flying overhead holds in its beak a liberty cap labeled "the Constitution must be preserved," c. 1834. Library of Congress, Prints and Photographs Division.

Norfolk as well as creating another near the Gulf of Mexico, an oddly ambiguous locale in the latter case since only Pensacola had some provisions for the Navy.[46] Despite its small size, Pensacola was vitally important to the Navy since it was best positioned to protect the Navy's strategic asset—the wide swath of live oak trees that had given the Navy a significant advantage over its earlier French and British opponents. Live oak, the board noted, was the object of pillage by unprincipled nations. It was one of the densest woods in the world, and its incorporation into the USS *Constitution*'s hull, for example, led to its moniker "Old Ironsides" when British cannonballs simply bounced off. Regarding live oak, the majority of funds spent between 1827 and 1830 were on the purchase of the live oak cut to molds. Other costs included purchasing land, agents' salaries, and examinations and surveys.

In addition, the board could justify a Navy yard in the Gulf of Mexico for economic security. Colonial and early republic era US trade depended heavily on maritime routes to the West Indies. This was one reason why Jackson expanded that squadron. By 1829, the Mississippi River system accounted for half of all US exports.[47] As Jackson learned during the Battle of New Orleans and the journey to his inauguration, the river system had become the lifeblood of the economy and ought to be viewed from a strategic standpoint. This viewpoint was not unique to Jackson and the Democrats. One of its strongest supporters in the senate was Samuel Southard. The former secretary of the Navy under the second Adams now served on the Committee on Naval Affairs. Like his colleagues, he supported deepening Pensacola Bay, constructing a dry dock there, and building a marine railway for the repair of warships.

As was and is standard in military funding discussions by presidents or members of Congress, local politicians and interests advocated for their communities. Jackson's presidency was no different. For example, in 1836 the mayor and city council of Baltimore appealed to the president to establish a Navy yard there.[48] Similarly, two years earlier, community business and political leaders in Charleston advocated a naval depot and dry dock in that key South Carolinian city. Senator William J. Grayson, also serving on the Committee on Naval Affairs argued on behalf of the capacity of Charleston's harbor, the security it provided, and the harbor's depth. He also added it had access to the best timber for shipbuilding, given that it was closer than any other major yard to the live oak plantations. Various advantages would result from establishing a Navy yard in that city for the building of smaller vessels of war as well as frigates, a process rendered easier with the assistance of steam.[49] To support his argument regarding the depth of the harbor for large, deep-draft ships, Grayson argued that during the Revolutionary War, two British 52-gun ships of the line had passed over its bar safely. It is unknown how Jackson, who had lost his mother due to those ships, received Grayson's appeal, but the Navy would not establish a yard in Charleston until decades later.

As part of the focus on Navy yard facilities, Jackson was intent on continuing the previous administration's policies constructing dry docks. The cost of dry docks, like ships, was largely a function of material and labor. Dry docks were constructed of granite. The first dry dock was built in Norfolk (fig. 4.5) and was considered "one of the most important feats of engineering to be undertaken in America during the pre-railroad era."[50] Its construction had begun in 1827 and was completed in 1834 at a cost of $950,000, which was an additional $300,000 of the initial project. A second dry dock was constructed in Boston. Not surprisingly, Jackson took clear, direct interest in the Norfolk dry dock. In December 1830 Jackson, writing in the third person, made his intent known to Branch: "The President desires that you should make known his solicitude to have this Dock compleated [*sic*] as early as possible, that our large vessels of war may be got in them for repair & prevented from that ruin that must ensure from their present condition."[51]

4.5. USS *Delaware* in the Norfolk Dry Dock. Norfolk's Dry Dock #1 was the first built in the Navy and an important step for the fleet. Courtesy of the Hampton Roads Naval Museum.

Two of Jackson's phrases in this order shed unexpected light on his policies regarding the Navy. The first was for the rapid completion of the dry dock. The dry dock was unprecedented in the American republic and was a significant investment, which proves again Jackson's support for the Navy or, at least, for the industrial capacity, should a future wartime need arise. The second phrase is that it be completed "as early as possible." A more conservative and political approach would have been to extend the work on the dry dock, thus providing jobs for the community for a longer timeframe. Instead, Jackson's desire for speed demonstrates a desire for financial efficiency and a desire that older ships would not be broken up or serve as receiving ships. Jackson meant to prepare the fleet fully for war should the need arise. By 1831, the dry docks were nearly complete. According to Secretary Woodbury, they "present[ed] to the eye specimens of stone masonry seldom rivalled in beauty and solidarity."[52] In 1835, Woodbury recommended a third dry dock, this one to be constructed in New York. The Brooklyn Dry Dock, however, would not reach completion until 1851. Two dry docks seemed enough for the Navy's size.

By 1831, Jackson had effectively assessed the immediate and long-term needs of the Navy. Ships that required maintenance and not deemed unfit received an infusion of funds. Partially completed ships at various stages of construction likewise received funding to ensure they did not rot in the shipyards. Just a year after investigating the issue of yard consolidation, the administration reconsidered its initial view. Branch reported to Congress that it should leave open the possibility of more, not fewer, Navy yards. It is possible that this was done for political expediency. For example, the proposal for a Charleston Navy yard (in 1834) came a year after the Nullification Crisis.[53] Had it been acted upon, it might have curried favor with the local politicians and citizenry. Moreover, given Jackson's own statements on the future of the Navy, he foresaw a real need for expansion. Most likely, the recommendation was based on both factors. Regardless, his administration argued early in his first term that additional yards would be helpful for repairs if smaller ships were built to increase the size of the Navy. That increase would occur in the middle of his first term. Having the infrastructure in place enabled Jackson to repair, maintain, and construct his Navy.

Return of the Schooners

As US trade expanded and international threats grew, Jackson authorized the most significant buildup of the Navy since the height of the War of 1812. In 1831, his administration built or acquired eight ships, more than in any other single year since 1814, when Madison built nine.[54] Jackson also understood the value of smaller ships. They were less expensive to build, less costly to man, and deployed at a higher rate that than ships of the line or frigates. Aside from the War of 1812, no class of ship would play a more dominant and needed role than the unrated ships—sloops and especially schooners.

An act of Congress on February 3, 1831, authorized three new schooners: *Experiment* in Washington, *Enterprise* in New York, and *Boxer* in Charlestown, Massachusetts. This division of labor continued a long-standing expectation from members of Congress. The Navy Act of 1794 had authorized six frigates to be designed by Joshua Humphreys. Humphreys rightfully preferred the same shipyard build them. Doing so would ensure that he could supervise the construction of the super frigates that he had designed and that the shipwrights and construction workers would become more experienced with every warship they built. In addition, it would have resulted in more cost efficiencies since there would be a direct line from the resources to only one manufacturer (the selected shipyard). Distribution would be cheaper, for example, because it only required one route to the shipyard rather than multiple routes to multiple shipyards thereby reducing the need for extra personnel. Unfortunately, that was not what occurred in the 1790s, 1830s, or even through the early twenty-first century.

Since the construction of the nation's first six frigates, the work was distributed to as many congressional districts as possible as were the Jefferson gunboats a decade later.[55] This was regardless of the overall cost of the program. When government funds were at stake, it was advantageous to gain the support of as many members of Congress as possible. Such support could only be enabled by distributing funds—and thereby providing jobs—among several congressional districts. For Humphreys and his successors in the 1830s, including his son, this meant that ships were built to different standards and that some would simply be better constructed than others.

Of the eight ships authorized, built, or acquired in 1831, only one was a frigate and another a sloop. On February 3, Congress authorized ship construction. Sister ships *Boxer* and *Enterprise* were 10-gun, 194-ton ships with a length of 88 feet. Both designs by Samuel Humphreys were launched by the end of the year and commissioned the following year. *Experiment*, the only ship of her class, also mounted 10 guns but was slightly smaller at 176 tons. The William Annesley–designed schooner was launched and commissioned the following year.[56]

These ships represented both the reality of deployment rates as well as dangers associated with seafaring. In the case of the latter, between 1798 and

1860, twenty US Navy warships were either wrecked or disappeared. Two ships were wrecked in 1823 alone—*Enterprise* and *Alligator*. On September 11, 1829, the famed War of 1812 sloop *Hornet* was lost with all hands in a gale off Tampico, Mexico. Among her officers was Lieutenant Daniel H. Mackay, who had a lengthy correspondence with Lieutenant Samuel Du Pont. Their former shipmate Surgeon John S. Wiley recounted to Du Pont: "Our inestimable friend Mackay's bright career is closed forever, as all hopes of the safety of the *Hornet* have died away. . . . I feel the loss of our friend more sensibly than I expected or rather my uneasiness has been kept up day by day for several weeks until by meditating on the subject my imagination has painted him in all the dreadful agonies of the death struggle, and my flesh creeps at the horrid image."[57]

Perhaps there was more lost than friendship that day or the loss of the ship and its crew. Mackay's correspondence on political and naval policy issues reflect an unusual breadth of curiosity. More important was his intention to write articles about naval policy and affairs. Indeed, one letter suggests that he had already written one pamphlet under a pseudonym.[58] For 1829, this was a radical departure from the conservative, tested stance of junior officers to obey their senior leadership without question and certainly not to articulate them in a publication.

Mackay's correspondence with Du Pont is reflective of the curiosity and commentary given rise by a new generation of US Navy officers. Mackay, who received his commission during the War of 1812, could be as critical of his captain and the Navy as he could in expressing his commitment to it, even when he prepared to get underway, "not I confess without some foreboding, not exactly of death and yellow fever, but it's a climate I do not like."[59] A self-proclaimed "Jackson man," he suggested that Secretary of the Navy "Sammy Southard" "had the arrogance and presumption to take some officers to task for voting for the General."[60] There were the standard reports among friends of midshipmen returning to the ship "so drunk he could not stand and . . . fell overboard," regional conditions and operations, advice on women, commentary of books about steam defense, and politics at home.[61] While in the Caribbean, he reported on acts of piracy and how the officer manning a gun had his legs crushed and amputated during an engagement with one pirate ship.[62] By August, Mackay and *Hornet* had diverted from their intended visit to Havana to the affairs of Mexico. They had received reports about a Spanish force of several thousand soldiers under Brigadier Isidro Barradas conveyed by Admiral Angel Laborde.[63] Within weeks, General Santa Anna would defeat Barradas, and the *Hornet* would be lost with all hands off Tampico.

It is possible that the Navy's own era of enlightenment might have arrived a few years sooner than it did (as discussed in a later chapter). Or, perhaps, Mackay's loss encouraged his fellow officers to reflect on their Navy in a way that might not have happened without his existence and death. It certainly did not escape the attention of his friend, Samuel Du Pont, who was to have a significant role in the development of the Navy in the nineteenth century during two wars. Du Pont's own recognition and acceptance of Mackay's death was sufficient to be entered in his private journal: "I have been somewhat prepared while in Mahon to hear the confirmation of a most melancholy piece of intelligence—the loss of the Sloop of War *Hornet*. By the papers received here, the most sanguine must abandon all hope. . . . I have experience inward and deep sorrow for one in particular who has gone down with that decorated Ship."[64]

This may have had more of an impact on Du Pont and the rest of the Navy, given other incidents that year. In June, the USS *Demologos*—the first steam warship—exploded in New York due to an accident. On September 8, Du Pont's own sloop, *Ontario*, was en route to the Mediterranean when the weather that morning had a threatening appearance. The clouds banked up and a fresh wind on the quarter drove the sloop at nine knots per hour. By midafternoon, a gale rapidly approached the ship.[65] According to eyewitnesses, the ship was nearly lost. Had that been the case, in a period of three months the US Navy would have lost three of its thirty-eight commissioned warships.

All the ships lost were either sloops or schooners. This is not to say that larger ships were not at risk. While rounding Cape Horn in a storm on its return from a two-year deployment in 1824, the ship of the line USS *Franklin* came within two hundred yards of hitting rocks. Only the captain's quick actions saved the seven hundred lives aboard.[66] Still, it was the sloops and schooners that bore the brunt of the dangers as Jackson learned again in 1831 with the disappearance of the *Sylph* and in 1833 the wreck of the *Porpoise*. One might argue that smaller ships were more likely to be lost in a storm simply because of their size. Another cause might be the ship's leadership. As unrated ships, they did not have captains in command but rather lieutenants or masters commandant who had less experience. There is also a third possibility and that is the level of deployment.

Sloops and schooners were the backbone of the fleet performing duties throughout the world. They were the ships that could most quickly be ready for overseas deployments including assembling crews. Ships of the line represented diplomatic authority and perceived national strength since few countries could build or maintain them. Frigates likewise performed these roles, particularly the ships like *United States*, *Constellation*, and *Constitution*. Both classes served admirably as flagships but were too few and too manpower intensive to deploy extensively. Although a full study of available ship logs remains to be complete, one way of understanding the usage of various classes is to study the fleet's location at a given time in the year. The best source for that, as previously suggested, are the annual reports of the secretary of the Navy.

Those reports provided the condition and location of the fleet. Based on that available data, a rough idea about deployment rates emerges, although the numbers are somewhat inflated. A rough estimate for the average annual deployment rate for ships of the line from 1825 to 1835 was less than 10 percent; in the first term of Jackson's presidency no ship of the line deployed (fig. 4.6). The average annual deployment rate of frigates during that period was 37 percent. More telling was that the deployment rate under the second Adams administration was 47.2 percent; during Jackson's presidency, that rate fell to 28.7 percent. By contrast, sloops had an average annual deployment rate of 68.2 percent; schooners, 76.6 percent. Consequently, the greatest need was for the smaller ships.

Jackson also understood that the unrated ships served an important role in maritime security along the US coastline. The US shipyards were at capacity in 1831 with the construction of five ships. Jackson also had a need to protect the strategic resource of live oak plantations in the Gulf of Mexico. When reports of European smuggling through intermediaries of the trees for their own shipbuilding programs reached Jackson, he took immediate action.

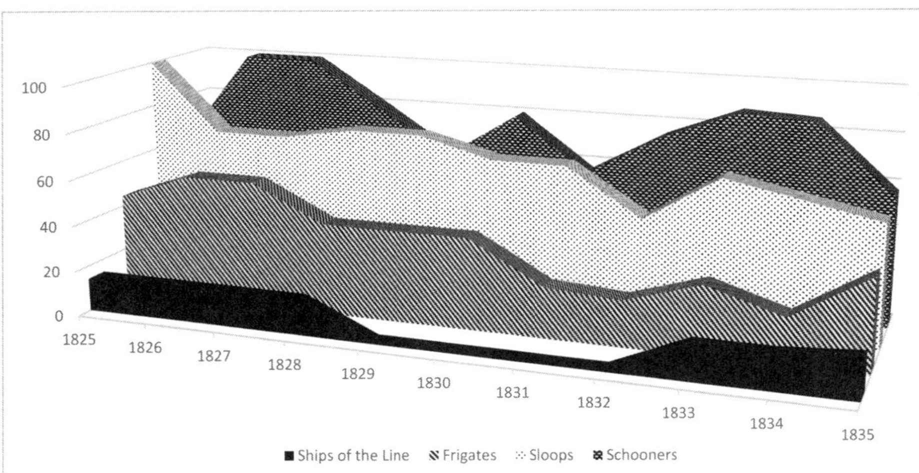

4.6. Ship deployment rates, 1825–35. Compiled by author from Annual Reports of the Secretary of the Navy.

He ordered the purchase of three schooners in Baltimore, the *Spark*, *Ariel*, and *Sylph*, for immediate deployment to the gulf. These were lightly armed (reportedly one-gun), fifty-ton ships.

It is unlikely that any of these ships had a real impact on protecting the wood from poachers. Within a year, the *Sylph* had disappeared, and both the *Spark* and *Ariel* returned and were sold. Regardless of the losses, Jackson's support of a growing Navy in 1831 is yet another piece of evidence that suggests his navalist leanings. To secure his position in that camp, however, required more than just building smaller vessels. Navalists supported ships of the line. Jackson fulfilled that in his second term with one of the largest ships of the line built by any country, the USS *Pennsylvania*. To do so, the Navy needed a designer capable of doing so.

The Designers

Upon first reflection, Jackson's naval policy was strictly in line with the anti-navalists, but it was far more pragmatic than ideological. It simply made no sense to the administration to continue drawing from the strategic reserve of live oak only to leave them to the elements in the shipyards. Temporarily stopping new authorizations of ships would allow the administration and the yards to complete construction of the already authorized ships. For this, Branch relied on the British system of using a select few designers and shipbuilders. Ship designers and even naval officers went to England to overtly or covertly study British ship designs, though the American Navy was just as likely not to adopt them. This was the case in the 1830s when the Americans could have learned from Sir Robert Seppings.

Seppings served as England's surveyor of the Navy from 1813 until 1832. Although Joshua Humphreys had radically improved the US frigates prior to the turn of the century, American ship innovation was largely stagnant unlike the British. Seppings made three major contributions to ship design that the United States failed to discover or adopt.[67]

First, he modified the bow from a beak-head bulkhead to rounded bow. The former proved a costly design. Though acceptable during traditional lines of battle in which fleets fired broadsides in parallel to each other, the Battle of Trafalgar proved that the beak-head bulkheads were too thin and weak to sustain heavy fire. This was especially true of Nelson's flagship, HMS *Victory*, whose gun crews in the fo'c'sle suffered the greatest casualties as it sailed directly toward the Franco-Spanish line.

Seppings's second contribution was similar to the first as he created the curved stern bringing the frame timbers to the upper deck. The stern was traditionally the weakest part of the ship since the quarter galleys for the commanding officer and admiral were tacked on to the hull and supported by a single horizontal crossbeam. Seppings offered the curved stern, which mitigated the stress but also enabled a ship to mount more guns especially on the stern and port quarter, always the most vulnerable locations on the ship. Despite the improvement, Seppings's design faced significant resistance from the admiralty and other shipbuilders as they deemed the design unattractive. Seppings wrote to the Royal Academy of Science to defend his position.

His third contribution was to support the ship from the interior with diagonal frames instead of the ninety-degree riders in the American frigates. Again, speaking to the Royal Academy, he demonstrated the strength of a triangle compared to a square and how that would impact the interior of a ship. The impact of these three innovations suggests that the British models by 1830 were superior to those of the outdated American ships of the line. Despite this, the reform government relieved Seppings in 1832. It deemed the Navy board an antiquated and inherently corrupt organization, and it replaced civilians with flag officers. The surveyor was the only position to survive, but King William IV favored William Symonds, a former naval officer and yacht designer interested more in ships' speed than durability and firepower.[68]

American ship architects and builders were too few to form a formal school like England's School of Naval Architecture that served under Seppings and instead relied on the traditional mentor system.[69] The father of American naval architecture was Joshua Humphreys, whose uniquely designed super frigates enabled the United States to gain single-ship action victories against the French and British in the Quasi-War and War of 1812 respectively.

The role of chief naval constructor, based in Philadelphia, fell to his son Samuel Humphreys, who served in that position from 1826 for the next twenty years. Captain William Bainbridge, at that time serving as president of the Board of Naval Commissioners, officially endorsed Humphreys. Bainbridge commended Humphreys as advantageously known to the public and particularly to the commissioners. The captain considered him one of the most distinguished naval architects in the country and eminently suitable for the position.[70]

Along with Humphreys, the Navy hired John Lenthall the following year to serve as Humphreys's apprentice. By 1828, Lenthall became an assistant naval constructor at the Philadelphia Navy Yard and continued in increasing positions of responsibility through 1871. Jackson's first term also included the services of veteran shipbuilder Henry Eckford. Eckford's construction of ships in the Great Lakes had significantly contributed to halting Britain's advance from Canada during the War of 1812. He may also have had more influence with the administration than even Humphreys himself. In June 1829, Eckford provided extensive recommendations to Branch on changes to the Department of the Navy. He proposed more control over the Navy by civilians instead of naval officers (such as the Board of Navy Commissioners). Eckford complained that naval officers criticized the labors of the civilians "who are not always competent to decide."[71] A peacetime Navy, he argued, demanded strict economy, but each constructor could improve the quality of ships. The present organization of the Department of the Navy was too defective to serve that purpose. Overall, he made a dozen recommendations including decreasing the ranks of naval officers and increasing civilian control of the shipyards.[72] Eckford had not initiated these recommendations but only responded to Branch, who had already surveyed the opinions of senior naval officers. Eckford returned to the private sector. In one project, he built a steam corvette—the *Kensington*—for the Mexican government, which did not accept it because it was too large. Eckford instead found a buyer in Russia, where the ship was renamed the *Prince of Warsaw*.[73]

When Congress funded the construction of the largest wooden warship built in the nation, it again turned to Humphreys. "I was aware of the great responsibility incurred," he later wrote, "and I was also aware that if any accident that happened in launching that ship, whatever professional reputation I might have had, would depart from me forever."[74] It was not just Humphreys's reputation at stake but the ability of the nation to construct its first line of battle ship in nearly a decade. The completion of the *Pennsylvania* would mean an increase in national prestige when it entered foreign ports. There was also a domestic reality in that it created jobs.

Pennsylvania was not a creation of the Jackson administration. The ship had been part of the 1816 authorization bill. What set her apart was she was the only first-rate ship of the line proposed in the US Navy. The design was for a 120-gun ship as opposed to the third-rate 74-gun ships of the line authorized that year. Her keel was laid six years later, but then work on it stopped. She would be ignored until Congress took up the 1837 naval appropriations bill just a few weeks before Jackson left office. The bill included a provision to fund *Pennsylvania*'s completion. When the provision was struck from the bill, Philadelphia congressman Joel Barlow Sutherland rose to defend the funding. The ship, he argued, needed to be built rather than left rotting under a roof.[75] He challenged the chair of the Naval Affairs Committee, Leonard

Jarvis, as "having expressed preference for small sloops and schooners."[76] One of the reasons for this line of argument was the fear that if a ship of the line were lost, it would take hundreds of American sailors down with it. This ship, arguably, was too big to fail. With a crew complement of approximately one thousand sailors, the ship would represent approximately 10 percent of the entire size of the naval force in terms of sailors. Sutherland proposed that the *Pennsylvania* ought to be fitted out and launched and sent to foreign ports, that it might there be seen what American naval architecture was, what the seamen were, and what force the country could command in war.

Congressman Jarvis counterargued that the ship was so large that only yards in the North could build it, so its construction unfairly favored northern states. Other members of Congress opposed the program because of its cost. Representative John Reed, who had served in Congress for twenty years—alternately as a Federalist, anti-Masonic, anti-Jacksonian, and Whig—proposed the money for *Pennsylvania* be shifted to domestic programs. Annual maintenance costs for the ship were estimated at $70,000. According to Reed, this was wasteful, given that five ships of the line were at yards. Reed made a valid case. From 1825 to 1835, ships of the line had a deployment rate of 17 percent, and for three years, the Jackson administration had deployed no ship of the line. If *Pennsylvania* were built, then its fate was likely to be found at American wharves rather than in serving overseas. Reed was ultimately proven correct.

The Committee on Naval Affairs denied funding for the *Pennsylvania* in committee. Rather, it supported funding more sloops and schooners. The $400,000 earmarked for *Pennsylvania* could instead construct eighteen sloops and schooners. This would have accomplished several goals. First, it would immediately expand the Navy's fleet. Second, doing so would increase the number of ships available for deployment. Third, it would have eased the burden on ships that were deployed for long periods, since sloops tended to have the highest rate of deployments. Fourth, sloops usually had a lieutenant or master commandant in command rather than a captain. This was important since advancement was slow since captains did not retire. Some captains had held the rank since before the War of 1812. More smaller ships meant more early command opportunities for those lieutenants or masters commandant who already had the maturity and experience to lead and to prepare them for greater assignments.[77] Fifth, having additional smaller ships meant more opportunities for midshipmen to learn their profession. Serving as one midshipman among two dozen on a ship of the line took longer to learn the intricacies of sailing a man-of-war. Serving on a sloop meant one midshipman among a few had comparatively more responsibilities and an abbreviated learning curve.

A time factor also intensified the *Pennsylvania* debate. Silt deposits were quickly increasing in the Delaware River. The ship, at a projected 3,200 tons, would be 50 percent larger than any other American ship of the line with a draft that would make it impossible to launch in the future. In addition, new street construction in Philadelphia near the shipyard would further complicate and potentially delay construction. When the full House voted, it supported the committee's recommendation by a vote of 125–55. Of those voting in support of the ship's construction was John Quincy Adams, who had been elected to the House after his defeat for the presidency.[78] Given the overwhelming rejection of the ship by the House, *Pennsylvania* should not have been constructed. Instead, the Senate quickly revived it. Records do not reveal the reason for this reversal. The chair of the Senate's Committee on Naval Affairs, William C. Rives, offered an amendment for $100,000 for launching the ship. After passing in the Senate, it went to the House, where it passed 91–87.[79] Only two days remained in Jackson's administration. There is no record of Jackson's official position, but given that he did not veto the bill, he gave its construction at least his tacit support. Supplemental appropriations were passed to complete the construction. The ship launched in July 1837. An estimated two hundred thousand Americans attended the ceremony.[80] James Fenimore Cooper, who attended the launch, wrote to his wife, "She is, altogether, the best-looking three-decker I have ever seen."[81] In November, *Pennsylvania* sailed to Norfolk. In its twenty-four-year life, it never received its full complement of sailors, it never received enough cannons for its design, and it never made it to the open ocean. At the outset of the Civil War, she was burned to avoid capture by the Confederates.

There was no real need for *Pennsylvania*'s construction. From an operational standpoint, Reed's evaluation was the most accurate. Enough ships of the line were already available for deployment if needed. *Pennsylvania* contributed nothing to the deployments of American squadrons since many squadrons deployed with only a frigate as the command ship. Technologically, this class of ship was already fading from the world's oceans. To make the class relevant, the Royal Navy's last sailing ship of the line, HMS *Queen*, was launched two years later and eventually configured with a steam engine. In fact, all Royal Navy ships of the line after 1839 were designed to include steam power. The last of the Royal Navy's mechanized wooden ships of the line was HMS *Victoria*, which launched in 1859 and was followed immediately by the ironclad HMS *Warrior*.

The *Pennsylvania* reflected a divided nation politically and strategically. It divided North from South, the latter suspicious of any federal funds for large projects in northern states. It divided the navalists between those who sought the prestige of large ships and those who saw the benefits of building more smaller ships that were more practical. In Congress, the naval theorists were cast aside in the perennial fight for federal funds for districts. The ship represented a half-year worth of hundreds of jobs in Philadelphia. Jobs meant votes for whatever member of Congress secured those funds. In the end, an expanded Navy was undermined by the need for reelection by members of Congress.

This factor may have also played a role in presidential elections. Jackson's second vice president, Martin Van Buren, won Pennsylvania in the 1836 election with only 51 percent of the vote. Work finished on the ship by mid-1837, just as the economic panic was underway. Had the administration and the members supported the construction of eighteen sloops and schooners—as originally proposed—that work would have extended through at least two years and several states. Van Buren would lose to William Henry Harrison in 1840; in Pennsylvania, Van Buren lost by the scantest of margins—50 percent to 49.9 percent—thus diminishing Jackson's immediate naval legacy with the transfer of power from his party to another. Nevertheless, it would be the last ship of the line built with the intention of deploying by the country as sail gave way to steam.[82]

Steam Comes to the Navy

The transition of sail to steam was gradual for several decades. It then grew rapidly both commercially and militarily in the 1840s. The most prominent Jackson biographer argues that "Jackson did not have any sense of what was coming, but it was coming. I don't think he had that kind of vision. He didn't bother to think about it."[83] However, Jackson was not only open to the use of steam technology for the Navy, he actively advanced it. Neither of his two predecessors built a steam warship, and only one purchased a commercial steamship for military operations. Jackson's administration ordered naval officers to gather intelligence on steam warship technology to inform domestic decision making on naval policy. He pursued the incorporation of steamships into the Navy's force structure and was the first president to use steamships in joint Army-Navy actions. Perhaps this was the most enduring of shipbuilding values in the American Navy as it was in other navies. Historian Bernard Brodie writes: "Of the several great naval revolutions of the nineteenth century, the introduction of the steam warship was in its tactical, strategic, and ultimate political consequences by far the most important. No other invention influenced so profoundly the conditions of naval warfare and the factors contributing to naval power. Yet in the two decades that followed its first appearance, the steam warship gained but scant attention. Its introduction was stubbornly resisted by naval officers and largely ignored by statesmen."[84]

This is only partially accurate. There is no question about the impact of steam power in the latter half of the nineteenth century. Americans had had some exposure to steam engines, but there had been little opportunity to fully exploit its widespread applications. Steam power was becoming more common by Jackson's administration but was still in its infancy as the United States did not yet have significant industrial capacity. Steam ferries, such as one first built by John Stevens in 1802, were inefficient but slowly improving. Despite this, officers saw the future. The prescient but doomed Lieutenant Daniel Mackay, who died with all hands aboard *Hornet* in 1829, foresaw steamships as the means of war at sea in the coming decades. Reviewing a book on steam defense and vulnerable points of Britain, he relayed to his friend Lieutenant Du Pont of the "absolutely necessity and great facility of protecting our Coast by Steamers . . . this now universal mover to the purposes of Naval War."[85]

The US Navy's inventory at the beginning of 1829 included two steamships. Neither was operational. The Navy in 1822 purchased the *Sea Gull*, built in 1818 as a commercial vessel in New England. David Porter, fighting pirates in the West Indies, required shallower draft vessels. The *Sea Gull* offered that as well as windless maneuverability in the littorals. In addition, the ship could make round trips back to the United States faster than most sailing vessels. In a two-year period, she made three deployments to serve with the West Indies Squadron. By 1825, the ship was considered unfit for further deployments and converted as a receiving ship in Philadelphia until 1840. *Sea Gull* was the first operational steam warship, but it was not the first steam warship in the Navy. *Sea Gull* was preceded by Robert Fulton's *Demologos* (fig. 4.7), the first steam-propelled warship in the world, but it never saw action. *Demologos* was recognized as a crude vessel since it was the first. Even proponents suggested it was "clumsy, cumbersome, and very slow in her movements."[86] Still, they viewed that the formidable character of the vessel was never actually tested.

Fulton had built *Demologos* in 1815. Although it was not completed in time to fight in the War of 1812, it still drew interest among officers who knew they had barely escaped a broader war loss and disaster at the hands of the British. It was the only steam-powered warship built for the next twenty

4.7. *Demologos* was built by Robert Fulton, but too late to fight in the War of 1812. It was destroyed in an accident in 1829. Courtesy of the US Naval Academy Museum.

years and even its accidental destruction in 1829 did not deter technology's advocates. Articles in the naval journals recognized that "the whole naval system of this and every other maritime country is about in a few years to be overturned by the superior powers and advantages of steam frigates, steam fortifications, and steam guns."[87] The change, they saw, was inevitable and US warships should be built or retrofitted with steam propulsion.

One or a few steam warships might be used in the West Indies for piracy suppression, but a few would not be enough in a confrontation with a greater power. With England's twenty-one steam warships in 1836 and France's twenty-three, a few American steam warships could quickly alter the disparity in naval capabilities between the two powers. This was not a new concept in American international relations theory. During the debate on the ratification of the proposed Constitution to replace the Articles of Confederation, Publius argued in Federalist 11 that "there can be no doubt that the continuance of the Union under an efficient government would put it in our power, at a period not very distant, to create a Navy which, if it could not vie with those of the great maritime powers, would at least be of respectable weight if thrown into the scale of either of two contending parties."[88] If that held true in the 1830s as it was proposed in the 1780s, then the United States itself could compete with larger countries by investing in its own steam warships. They could also be used in stabilizing the Gulf of Mexico, given the belligerence of both Texas and Mexico and the potential growth of piracy.

That honor fell to Fulton's *Demologos* upon which the Navy had been experimenting for more than a decade. She was, in fact, the first steam warship built by any Navy. The shallow-draft, twin-hulled ship constructed during the War of 1812 was intended to support coastal, riverine, and harbor defense rather than blue-water operations. Primitively—but understandably—fueled by burning wood, the ship was still an important genesis for the eventual coal-powered steamships of the later 1830s and throughout the nineteenth century.

By the time Jackson became president, *Demologos* was at the Brooklyn Navy Yard with a full crew including a captain and crew of more than thirty. Primarily serving as a receiving ship, she was moored head and stern opposite the Navy yard. According to the commanding officer in 1829, Captain John T. Newton, the ship was in such a decayed state that the decks leaked.[89] The guns were used sparingly out of fear they would go through the deck. Mounting sixteen guns, she had five hundred pounds of gunpowder partly in casks in the magazine located along the starboard bow. A standard door stood between the magazine and the light room.

On June 4, 1829, in addition to the crew, the ship had prisoners awaiting adjudication and more than seventy recruits. Newton was escorting Commodore Isaac Chauncey on board for an inspection that lasted until mid-afternoon. Chauncey and the captain left minutes before an explosion destroyed the ship, killing approximately thirty officers and sailors and wounding another twenty-nine. Any opponents of the comparatively new steam technology might have used it as an excuse to slow down the pursuit of this radically new type of ship. Branch constituted a court of inquiry to determine the cause. For three weeks, the court heard testimony from the surviving crew. The captain testified that his orders, especially regarding the magazine, were particular and verbal to all officers on the ship. The ship always had a sentinel placed at the hatch so as not to permit any improper intrusion. "Every precaution was taken at all times to prevent accidents."[90] Clearly, the captain was wrong in this case since the event had been the result of an accident.

Commodore Chauncey had no knowledge of the cause but stated that it could not have been the result of the design. It had to be, he stated, accidental. He speculated that either the gunner's mate entered the magazine with his lantern or he entered the passage to the light room where the atmospheric state—without ventilation—produced combustion by coming in contact with the lighted candle. According to Marine second lieutenant Alexander Mull,

Acting Gunner's Mate Thomas Williams had come to the officer of the deck, Lieutenant Breckenridge, to enter the magazine. Breckenridge, later killed in the explosion, ordered him to have the fire and lights put out and a sentinel put over the hatch leading to the passage to the magazine. Mull was in the wardroom when the explosion took place, and it was, therefore, impossible for him to state the cause. If the fault rested with Gunner's Mate Williams, it did not change the opinion of the witnesses. Chauncey and others noted his lengthy service and unblemished record. Sailing Master John Clough testified that a few days before the explosion, the former gunner and he conversed about his successor. Both recommended Williams.

Even though Williams appeared to have caused the explosion, the inquiry explored another culprit. Midshipman David McDougal noted that he saw Seaman Hannan go into the hatch at some point and the sentry did not resist. The court asked if he knew that Hannan was "a person under charge of Stealing and a Bad Man."[91] Though there were several questions about Hannan, all that was certain was that his mangled body and broken legs were found after a week in the water. The body was not bruised or lacerated and the hair was not singed. This suggested that Hannan was not in proximity to the explosion.

The court's opinion was that the explosion resulted from an accident and not due to the design of the ship. It concluded that Williams probably had a lighted candle in his hand and that there was no error on the part of Captain Newton or the officers. One should not underestimate the court of public opinion, subject as it is to emotion instead of facts, because the process and result could have been far different. In American naval history, naval accidents have often spiraled into a media feeding frenzy based on assumptions, preconceptions, and political desires. The most glaring example of this was the explosion of the USS *Maine* that was caused by a design flaw. The media immediately claimed Spain had caused the explosion. The first two American investigations supported this theory.[92] While there was no similar effort to lay blame on a foreign power, any opponents of steam power might have used the incident to delay the development of this new technology.

In 1830, engineer John Stevens found that it was time to reassess his views on future naval tactics using steam vessels. He wrote to Jackson that the Board of Navy Commissioners was doing nothing to advance steam technology or to explore how it might be employed in naval operations. The conservative board at the writing of his letter in March 1830 was comprised of its long-term president, Rodgers, Captain Lewis Warrington, and Captain Daniel Patterson (Commodore Charles Stewart would replace Warrington later that year). Stevens held the entire board in little regard. He suggested they all ought to be promoted to admiral—a rank that did not exist until 1862—and moved out of the way so that they would not be an impediment to progress.

Stevens challenged their "violent and inveterate" prejudices.[93] Naval reform was vital, but the commissioners opposed any measures. Steam would be integral to the next naval conflict, Stevens argued. "Wo[e] be to that naval power who shall then remain unprepared to meet it."[94] Finally, he shared with Jackson the report of a naval officer sent to France and England to evaluate their naval establishments.[95] This was nothing short of an intelligence mission. Both European countries were making significant progress on their own steam engines and would introduce steam warships into their naval inventories.

It is unknown how Jackson reacted to the letter. But it is evident that early in his administration, Jackson was made aware of the steam power issue through one of the few men in the country who had authority on the matter. Stevens was the inventor and engineer who had built the first domestic steamboat in 1802 and the first steam locomotive capable of towing passenger cars in 1825. In addition, Stevens carried with him the gravitas of one of the most pedigreed families in America. His wife was the sister of Robert Livingston, one of America's founders, and Senator Edward Livingston, who would soon serve as Jackson's secretary of state and minister to France. Jackson had seen the practical use of steamboats prior to the Battle of New Orleans and his inaugural journey, but Stevens now provided the commercial impetus for exploring this technology for the Navy.

Nevertheless, it is possible that Stevens overstepped his bounds among the naval establishment. A few weeks after he sent the scathing letter to Jackson, Stevens wrote to him again, promising "to be more guarded in my expressions on the present occasion."[96] If the Board of Navy Commissioners became aware of Stevens's scathing remarks, they could have made him a pariah in naval development. But Stevens was one of the few people in the country who could evaluate proposals for steam engines or bid on future contracts himself. The board also had to be careful of Stevens's family ties; having a powerful senator in the family tempered their response. Regardless, Stevens recognized the need to be more cautious in his approach on the issue.

Stevens's own history presaged the national industrial revolution that began to take hold in the 1830s in the United States. Before the American Revolution, the British had prohibited colonial production of weapons, such as muskets. It would take decades for the Americans to develop the ability to tool up and produce quality metals. The extraction of rich ores and other resources met rising indigenous manufacturing capability. This convergence immediately impacted the Navy. For example, in 1831 the Navy ordered the addition of iron tanks in ships of the line and frigates for both ballast and improved quality of water storage. This represented a departure from the norm as well an indication that the Navy and the country were finally able to produce the tanks.

The availability of basic resources, increased manufacturing capability, the creative vision by Stevens and others offered the Navy the first real opportunity to pursue a new generation of ships. While cultural resistance from some senior officers continued, the new possibilities had taken a firm hold in the officer corps. Master Commandant Alexander Slidell Mackenzie was part of the next generation to acknowledge the application of steam to warfare with specific respect to coastal steam batteries: "While such astonishing improvements are making in the application of steam, it would be improvident to overlook its probable importance hereafter in maritime warfare, or fail to keep pace with other naval nations in any new means of attack or defense."[97]

Secretary Woodbury would advocate for three such steam batteries in his 1833 report to Congress. Mackenzie noted the experiment of the *Demologos*. He recommended that Congress appropriate funds to the Department of the Navy to replace the *Demologos* or at least to purchase new and more appropriate machinery for test. Mackenzie's advocacy was not a meaningless entreaty from one of the many officers in the Navy. He was, instead, part of the Navy's royalty—the Perry-Rodgers clan that had dominated the Navy for decades. Mackenzie's brother-in-law was Matthew Perry (the younger brother of War of 1812 hero Oliver Hazard Perry). Mackenzie's own extensive authorship in the 1820s and 1830s and his membership in a naval dynasty were more likely to make members of Congress and Jackson listen. It also demonstrated a possible early split in the family given that the president of the Board of Navy Commissioners was Rodgers.

The resistance of senior naval officers to steam power was not universal. One of the most senior officers during Jackson's administration was Commodore Barron. He had had a checkered career. He was dismissed in 1807 from the Navy for five years for having surrendered the USS *Chesapeake* to the HMS *Leopard*. Prevented from fighting the British during the War of 1812, he was persuaded to settle a long-standing conflict with Stephen Decatur by duel in 1820 in which he killed the latter. Barron consequently served in command of several shore stations and shipyards including the Philadelphia Navy Yard—the center of American Navy shipbuilding. It was here that Barron would propose one of the Navy's most unique and innovative warships.

While commander of the Philadelphia Navy Yard, Barron had submitted several Navy-related inventions to the US Patent and Trademark Office. Some of his patents included a floating dry dock gun, a gun carriage, and a ship ventilation system. His most significant proposal, however, was for a steam-driven ram ship.[98] In December 1833, Barron reported to Woodbury that his proposed ship could be fundamental to the defense of America's inland waterways and coastline. He had discussed his design with other officers, none of whom dissuaded him from pursuing this concept.[99] Woodbury not only endorsed the proposal, he encouraged Barron to meet with members of Congress to advance the idea.[100]

Barron's ship had no guns; its sole offensive weapon would be its ram prow, much like ancient Greek and Roman galleys. He appealed to fellow Virginian Andrew Stevenson, at that time serving as Speaker of the House of Representatives. At Barron's urging, Stevenson had a model in the House Naval Affairs Committee room for display to help explain the concept to the members. Throughout 1834, Barron secured the support of more officers who saw the value in such a warship. The proposed ship would be capable of eight knots of speed, more than enough to pierce an enemy's hull.[101]

At the beginning of the next congressional session in 1835, the House Naval Affairs Committee supported construction of such a ship.[102] The committee noted the ship's simplicity and capability that went well beyond just being an experimental project. Already it envisioned a fleet of cost-effective, small, crew-size ram ships protecting America's coastline. With a proposed length of 250 feet, the ship was also unusual in that it would have three hulls. Between the hulls were steam-powered paddle wheels to protect them from enemy fire.

The Board of Naval Commissioners scrutinized the steamship. It had Samuel Humphreys and John Lenthall visit businesses that built steam vessels and engines in order to assess the Barron model. When the naval designer and constructor reported, the board concurred that the vessel "ought to be mainly constructed for harbor defense" travel at ten miles per hour and carry two heavy pivot guns able to throw eight- or ten-inch shells and four thirty-two-pounder carronades.[103] Humphreys and Lenthall also recommended the ship be about fifty feet shorter than proposed by Barron.

Despite the endorsement of the committee, the House failed to authorize funding. The lack of debate on the issue fails to shed light on the reason behind this, but there are four possibilities. First, Congress was concurrently considering the expensive ship of the line *Pennsylvania*. The House may have simply chosen to support one over the other, and since the *Pennsylvania* would result in more immediate jobs, that tended to be the politically expedient choice. Second, Barron's advocate in the House, Speaker Stevenson, had left office in the middle of his term in 1834, having been nominated by Jackson to serve as minister to England. Before Stevenson resigned, however, he was accused of having set into motion several of Jackson's policies, thus leading to accusations of a quid pro quo. This was ironic given Jackson's denunciation of the so-called corrupt bargain of the 1824 election. Jackson's opponents denied Stevenson his appointment. Therefore, it is possible the steam ram ship was simply wrapped up in this controversy. Third, it is also possible that Barron himself confronted a populist and hostile House that still remembered the death of America's greatest naval hero by Barron's hand, something he suspected when he approached Jackson himself.[104] Whatever the reason, the Naval Affairs Committee and Barron were prescient. The development of these ships in the 1830s would have had an important role twenty-five years later at the outbreak of the Civil War. Ramming of enemy ships came back into style, and the riverine systems were filled with makeshift steam ram ships. Barron's novel design benefiting from another generational development might have shortened the war. Regardless, the House did not abandon steam warships. A fourth and more likely possibility is that they chose to invest in a different steamship under consideration at the same time as the ram ship.

Fulton II

If Jackson seemed undecided about Barron's ram ship, he appeared to have endorsed another steam warship, which became known as the *Fulton II*.[105] The proponents of this ship had several advantages over the competing Barron design. The Board of Naval Commissioners invested more time inquiring about it, suggesting an openness that was not evident with the Barron design. It benefited from an intelligence-gathering mission by a member of the Rodgers family, while Rodgers served as the board's president. Finally, civilians experienced with steam engines were under consideration for contracts in a way that was absent in Barron's proposal.

In November 1833, the Navy sent Master Commandant Alexander Slidell Mackenzie to Britain.[106] Mackenzie published a two-volume book in 1835 about his time in Britain as if he were a common traveler observing architecture, people, and events, much like Alexis de Tocqueville did with *Democracy in America*. Absent from the book, however, was much mention of the Royal Navy for good reason.[107] Mackenzie had sent a full report on the Royal Navy's steam warship program directly to the Board of Naval Commissioners. Rodgers had ordered Mackenzie to go to England with a clear mission, to "procure as accurate information as possible relative to the construction of steam vessels of War in the British Navy."[108] When he reached England in early 1834, he visited the principal dockyards, where he observed the latest steamships and made acquaintances with Royal Navy officers. Whether this was spying or intelligence gathering, it was not unprecedented by other countries. One historian notes: "Two [French Navy] officers were sent to the United States in 1820 to report upon the properties of the steam vessels there, and this event was followed by the purchase and addition to the Navy List of a few unarmed steamships. But in the spring of 1830, six armed steamers were available as auxiliary warships and dispatch vessels in the expedition against Algeria."[109] Mackenzie reported that the British Navy had two types of steamships. The first was a smaller vessel for use as mail packets and as ship tenders. These ranged from 100 to 300 tons. The second class included offensive warfare ships at 600 to 900 tons and 147 feet long. Mackenzie assessed that their bows were sharp and could make 10 to 11 knots. He also gave specific assessments on individual British ships.

A broader value to the Mackenzie report was its intelligence on opinions of Royal Navy leadership and the inherent inability to get accurate information. Captain Sir William Symonds had been appointed surveyor of the Navy three years before Mackenzie's arrival. Symonds had proposed new modes of naval construction that enabled construction of larger ships, which met with opposition from traditional shipwrights and the School of Naval Architecture that had closed in 1832. Symonds's larger designs quickly became obsolete as steam power emerged.

Nevertheless, the two camps within the Royal Navy (the "old and new schools" as Mackenzie called them) were not reliable sources for the capabilities of the new steamships. Each side presented its own facts that best supported their case. For example, the new school tended to inflate ship speed and performance to gain the support of Parliament. The same was true of the sailing Navy who wanted to justify their own shipbuilding vision. Mackenzie's report was not solely focused on steam warships. Given the number of ships of sail that comprised the majority of the Royal Navy, Mackenzie also reported on newer ships such as the 50-gun HMS *Vernon*. Perhaps his most important observation regarding the traditional sailing Navy was how the Royal Navy was cutting down some of their smaller ships of the line into frigates such as HMS *Barham*. There was a direct correlation between British policies and American shipbuilding. It is no coincidence that the US Navy pursued the same with the ship of the line USS *Independence*, which was razed and became a 54-gun frigate in 1836.[110] In 1835, the secretary of the Navy requested $150,000 for the completion of the *Fulton II*. This was more evidence that Barron had little senior support for his design. Nevertheless, the Board of Naval Commissioners continued to be cautious about any option citing a lack of construction experience and untested cost estimates. The hull itself was already under construction when, with no decision made on the steam plant, the board advertised in major newspapers in the hope of contracting out for the more complex technology. The board was still unsure it could find any appropriate individual or firm, noting: "From [the board's] ignorance upon the subject of Steam Engines, they are in doubt whether the advertisement gives the necessary information to enable persons to make proper offers, and they are satisfied that they are incompetent themselves, and have no person under their direction which could furnish them with the necessary information to form a contract for Steam Engines . . . should the lowest offers happen to be made by persons whose general character and responsibility would not offer great security for their completing the Engines in the best manner."[111]

By January 27, 1836, the board had received only two offers. Gouverneur Kemble of the West Point Foundry proposed constructing the engine and boilers, while Daniel Large proposed manufacturing only the boilers. The design of *Fulton II* was changing and adapting to the latest improvements in technology. With the board paying generous attention to the subject, the members recognized that they lacked the expertise in the construction and management of steamships. Between the contractors, the board selected Kemble's West Point Foundry on February 17. Kemble had a long history with the Navy, having served as a Navy agent in Spain during the Second Barbary War where he learned about casting iron. He was also involved in Democratic politics in New York. After gaining the contract in 1836, he was elected to Congress.

One of Kemble's employees was Charles Haynes Haswell, a twenty-eight-year-old Marine engineer who had first worked on Robert Fulton's commercial steamship *Clermont*. Haswell was also connected with the Navy's reformers as one of the few civilians to hold a membership in the Naval Lyceum. Still, the board reserved full support for Haswell. Several firms and individuals sent letters of recommendation that concurred on his intelligence and familiarity with construction of engines. Nevertheless, the board remained concerned that he had no practical experience on ships. Despite this, Haswell was appointed first chief engineer of the Navy two days later.

By August, Rodgers reported to Acting Navy Secretary John Boyle that work on the *Fulton II* had been long delayed and that several more months were required to build the engines. Delays in ship construction, particularly of a new class, have been common in US naval history. It took another year, but the *Fulton* was finally launched in 1837, after Jackson left office. Even had there been a national security crisis or a sense of national urgency, it is unlikely that the ship could have been constructed more quickly. The industrial capacity of the country and the Navy were not yet mature, and, with this ship, the Navy could not get it wrong.

Captain Isaac Chauncey, having assumed the presidency of the board after Rodgers fell gravely ill, reported to Secretary Dickerson that the *Fulton* would probably be ready to accompany the ship of the line *Pennsylvania* if deemed necessary. Had that happened, it would have been an impressive sight, the new escorting the old. That might have been America's moment like what British artist J. M. W. Turner had captured with *The Fighting* Temeraire. Support for new steam warships was not universal—some older, conservative officers and later Secretary of the Navy James Kirke Paulding had reservations. However, they failed to collectively articulate their concerns in the pages of the journals whether by intent or necessity. One opponent did note: "As for steam, it will not do at sea, contending with ships of the line . . . this mode of warfare does not comport with the dignity of our nation. We have abundance of material for shipbuilding, skilful [*sic*] hands to construct, talents to direct, and bold hearts to use them."[112]

Jackson would not have fallen into the category of the conservatives. While the *Fulton* was under construction, the president had drawn the military into the Second Seminole War. Six months after the war started, Jackson personally ordered the deployment of three steamboats for riverine operations in Florida.

Jackson's presidential term was more navalist than antinavalist given the number of ships he authorized to be constructed. But Jackson was not only interested in raw ship numbers. His administration reflected a mature understanding of national needs for the fleet and the types of ships that best supported the various missions required. He displayed a remarkable balance of investing in the highly deployed sloops and schooners, the prestige offered by one of the world's largest ships of the line, and the advancement associated with supporting new steam technology. In addition, he authorized the construction of the first armed supply ship, a radical departure from the first six presidents. While it was only one ship, it was a clear message that the US Navy would be expected to sail to any distant station and support itself if necessary. He understood the importance of investing in shipyard infrastructure, an important albeit overlooked aspect of advancing a nation's Navy. Because of Jackson's administration, the Navy could build upon fleet diversity and infrastructure support to conduct more global operations.

5

Maritime Destiny—A Global Strategy and Naval Operations

The United States was not at war with any country during Jackson's administration. That did not, however, mean threats to the nation were dormant. Jackson's and white southerners' views of Native Americans as threats led to the Second Seminole War in 1835. In reality, the Seminoles—and other Native American tribes—posed no existential threat to the United States like a European state might. The threat, instead, was perceived and amplified by the Jackson administration and settlers. The war was initiated by the Jackson administration. Nevertheless, civil wars, maritime threats by pirates, and suspicions about the great powers were always on the near horizon. Jackson himself recognized in his inaugural address and annual messages to Congress the various challenges the young republic faced internationally. He understood that those global challenges threatened US commerce and economic growth and developed a national maritime strategy that planned for and addressed them.

The components of Jackson's strategy were more inspired than historians have generally contended and have been largely ignored in the historiography of the era of the early republic. Still, a few US naval historians have expressed a belief that Jackson did reflect on the issue, but they failed to study his administration in depth. In a book on naval history, Harold Sprout and Margaret Sprout wrote, "It is possible that [Jackson] had some conception of the strategic doctrine which is summed up in the phrase, command of the sea. If so, he had clearly outstripped even the professional naval opinion of the day."[1] Another historian suggested that under Jackson, "any nation, large or small, challenging American rights on the high seas would find itself on the dueling ground looking down the barrel of the Tennessean's pistol."[2] The reality is that Jackson apparently understood the limitations of his Navy against a larger European power and, in the cases of smaller foes, had a mixed reaction.

Jackson's maritime strategy was comprised of three parts. The first was that he would create favorable trade conditions for US merchant ships. Arguably, his opening of additional commercial markets in the 1830s was more valuable to the United States than Mahanian expansionism of the 1890s. In the case of the latter, Alfred Thayer Mahan, Theodore Roosevelt, and others wanted territories in the Pacific in order to facilitate naval operations in future wars. It also demanded a large, costly fleet. But it was the peaceful pursuit of trade agreements by John Quincy Adams and Jackson that helped to fuel the industrial revolution that enabled the Mahanians to believe American dominance on the high seas was possible.

The second component of his maritime strategy was the continued provision and expansion of squadrons at key geographical locations to protect the merchant ships. The Navy, Jackson said, was the best standing security of the country against foreign aggression. It was also the only way to ensure that the lawful transaction of maritime commerce continued unabated. Without security, American merchant ships were a target of pirates and other interests. This meant insurance rates rose, which drove up the cost of conducting business. That increase in cost would then be passed on to the American consumer.

The third part of Jackson's maritime strategy required punitive strikes if necessary. Contrary to conventional historical assessments, Jackson did not use the Navy without significant thought on the options, approaches, and consequences. For Jackson, the third option was always a last resort, an important distinction to dispel the myths about Jackson's temperament with regard to international affairs. First strikes in naval conflicts were contrary to his philosophy. "Our country," he told Congress, "is not in a situation to invite aggression, and it will be our fault if she ever becomes so."[3]

The Geostrategic Situation

Jackson understood the unique position of the United States. To the country's north was a largely peaceful border with Britain's territory of Canada, though the official boundary would be in dispute until the Webster-Ashburton Treaty in 1842.[4] To the west lay lands nearly every American sought in the pursuit of growing the country. Mile by mile, Americans seized native tribal lands in pursuit of national boundaries that would eventually stretch from the Atlantic to the Pacific. To the southwest lay the Gulf of Mexico, and to the east, the Atlantic. America's location separated the country with what Jackson often called "wide seas."[5] The Atlantic meant that America was far removed from the major powers that could threaten its coast. This had been the case during the American Revolution and the War of 1812, with Britain having trouble in sustaining naval blockades or invading armies. The Navy, Jackson said, ought to be a cherished arm of national defense. An ever-ready Navy could prevent or repel an attack. Absent large-scale invasions, the country had most to fear from attacks on its overseas commerce from abroad and harassment along the coastline. Oceans provided a natural means of defense for the country. Conversely, American merchants found themselves threatened by the expansive oceans and the myriad of threats, both natural and man-made. Following the War of 1812, the United States began to establish squadrons in key regions of the world that would allow both protective operations and quicker punitive actions if necessary. The first squadron was deployed to the Mediterranean, largely to counter the Barbary States. Three different squadrons had been sent to Tripoli between 1800 and 1805 for the protection of American commerce and two more in 1815 to Algiers. The nation began investing in permanent stations rather than ad hoc squadrons.

Other squadrons followed in the Monroe and second Adams administrations. In 1821, the Pacific Station deployed to protect American whaling ships. The West Indies Squadron followed in 1822 to repel pirates and a Brazil Squadron in 1826 in the middle of destabilizing civil wars in South America. By the end of Jackson's second term, he would establish the second to last of the overseas squadrons as he expanded trade in the Far East with the East India Squadron.[6] In a matter of less than two decades, the United States had positioned itself globally. The squadron system meant that America was a slowly emerging world power. Unable to conduct large fleet actions, like the Battle of Trafalgar, even its small squadrons represented a large investment for the country. But the squadrons offered immediate and long-term benefits. In the immediate term, they offered protection for regional American commerce. In the long-term, it meant that the small US Navy learned how to conduct extensive overseas deployments. This was no mean feat for any Navy. It required not only money but also responsibility for the squadron and ship commanders who operated in a near vacuum of information given the length of time it took for messages from the president or the secretary of the Navy to reach them. These squadrons were the basis for America's operational capability.[7]

Jackson had a history of belligerence throughout his military career against the British in wartime and against Indians in operations designed to expand American geographical boundaries. As president, he recognized that attacking Native Americans or pirates was more acceptable and less dangerous. This advanced the interests of his fellow countrymen who wanted to push westward for more land. Native Americans were recognized as savages and therefore received little legal protection. If a nation attacked individuals or groups that were deemed as uncivilized, international moral outrage was absent. The same reasoning allowed European empires to absorb wide swaths of Africa. Pirates were also easy fodder for nation-states to attack since they were criminals. It was quite different as president, however, for Jackson to attack another nation, particularly a major European power and, with it, the military forces they could bring to bear. Chaplain Thomas Lambert, aboard the frigate USS *United States* in 1836, found Malta to be a desirable place to visit and see catacombs or where St. Paul was reported shipwrecked. Leaving more of an impression, however, was seeing a British squadron at anchor—ten or twelve ships of the line. That one British squadron had more ships of the line than in the entire US Navy.[8] That fact was also not lost on Jackson, who, throughout his administration, had the strength of European navies reported back to him.

If he wanted agreements with Indian tribes, he could always overpower them militarily. Dealing with a nation-state, however, meant engaging in real diplomacy. Jackson's public statements about other countries were quite different from those he had in private correspondence. In his first annual message to Congress in December 1829, he declared that the United States was "at peace with all mankind."[9] He was laudatory toward most countries in his annual messages. His worldview was sophisticated, and his perspectives articulated in his inaugural addresses and annual messages to Congress on the importance of the flow of goods across oceans depended on the stability of international relations, goodwill between countries, and the ability to protect trade on the high seas. That trade required a Navy. As one naval historian notes: "The variety and wide scope of active naval service during this era of expansion in overseas trade had borne striking proof of the economic value of our growing Navy during intervals of peace. The pacific influence of naval force had been skillfully used to preserve our commercial interests and rights when as neutrals our trading ships were unduly hampered by hostilities between foreign countries and blockades incident thereto, to cultivate good relations which would promote commerce and to make treaties to the same end. When necessary, force had been effectively used to chastise backward peoples who had injured our property and taken American lives."[10]

In the brief inaugural address of 1829, Jackson said nothing about the international community, perhaps because he wished to wait until he was privy to communiqués between his administration and other nations. By the time of his first annual message to Congress on December 8, 1829, he was no longer the neophyte president. Events and conditions around the world shaped his international outlook. He was not an inward-looking executive only concerned with domestic policies. Admittedly, Jackson was prone to an emotionalism that overshadowed his domestic governance and politics, such as with the Eaton Affair and the Bank of the United States debate. He was a different person, however, when it came to international diplomacy. He was politically astute enough to know that, when it came to domestic affairs, his popular vote and position as chief executive of the nation allowed him some latitude in bullying political opponents. In dealing with Indian tribes, Jackson knew that any agreement he signed with them was meaningless, as he did not believe they could effectively defend themselves against the US Army.

His international approach, however, was one that rested on an understanding that he was negotiating with states, particularly European, that he could neither bully nor easily defeat. Jackson as a head of state, therefore, was thoughtful and measured in his communiqués with other heads of state. In each of his eight messages to Congress and his second inaugural address, Jackson was largely laudatory and conciliatory toward the great powers of Europe. This is not to say, however, that he failed to address the differences with other nations, which required resolution. "Our foreign relations . . . present subjects of differences between us and other powers of deep interest."[11] Nevertheless, his messages lacked the inflammatory language that he was more likely to use in domestic politics. But one historian would not forget that Jackson had a "fierce desire to win global recognition for the independence, sovereignty, and rights of the United States . . . [and had a] conciliatory approach with a willingness to accept practical solutions to problems in foreign affairs."[12]

What follows in this chapter is a picture of the world within the context of the work that is about the Navy. Each region, each squadron, each country and the diplomatic relations with each are worthy of a book-length work in themselves. Consequently, in the interest of and focus on the Navy during Jackson's presidency, readers are encouraged to also explore the vast literature available for those specialties.

The young republic's greatest naval experiences had been in the Mediterranean. At the turn of the century, its first squadrons were deployed there to fight in response to piracy threats from the Barbary Coast. The First Barbary War had established the Navy as a proven force capable of extended wartime operations thousands of miles from home. The conflict found junior commanding officers like Isaac Hull, David Porter, Charles Stewart, and others who remembered it as they served as captains and commodores over the next two decades. The Second Barbary War resulted in a permanent Mediterranean Squadron established and based at Port Mahon on the Spanish island of Minorca.

At the time of Jackson's inauguration, Commodore William Crane commanded the Mediterranean Squadron. It consisted of the ship of the line USS *Delaware* (74 guns); the frigate USS *Java* (44 guns); the sloops USS *Lexington*, USS *Warren*, USS *Fairfield* (all 18 guns); and schooner USS *Porpoise* (12 guns). By the end of the first year, the *Porpoise* had returned because it required repairs. Jackson's administration also recalled the massive *Delaware* because of his assessment of the political situation, which "did not require the employment of a ship of this class in that sea."[13] Frigates and smaller ships were deemed more efficient. To replace the *Delaware*, Jackson sent the frigates *Ontario* and *Constellation*, the latter more than thirty years old and a veteran of four wars.

Personnel problems permeated the squadron. Of the seventy-three courts-martial in the Navy in 1829 and 1830, thirty were because of incidents in the Mediterranean Squadron. Many were a result of drunkenness in one of the ports with the result that all the midshipmen involved were cashiered, actions approved by Jackson.[14] In another incident, a French lieutenant and an American sailor were killed in Port Mahon. It may have been this case the French minister of foreign affairs referenced when he wrote to the American minister at Paris to "call its attention to the necessity of establishing among the sailors on board of its vessels a rigid discipline, which may prevent the recurrence of such outrages as those of which Mahon has been the theatre."[15]

Crane charged two of his midshipmen aboard USS *Java* for leaving the ship without permission and disobeying the lawful order of their superior officer, Captain John Downes. The court found the midshipmen guilty and dismissed them, but Crane had to take additional measures since their actions had disgraced not only the individuals but all officers in foreign ports.[16] The Mediterranean was arguably the Navy's most crucial squadron given its direct relations with the most powerful empires in the world.

Since the founding of the country, arguably no powers held more importance in American policy than Britain, France, and Spain. "Of the unsettled matters between the US and other powers," Jackson wrote in his first annual message to Congress, "the most prominent are those with England, France and Spain. We have the best reason to hope for a satisfactory adjustment of existing differences."[17] Jackson's "best reason to hope" was the language of diplomacy. America was, at the time, a second-rate or even third-rate power. Since Jackson had little ability to force or even enforce an agreement, all he could do was hope. The first challenge was the potential volatility of Europe.

If Jackson could hold animosity toward any nation, it should have been Britain. His mother had died on a British ship, the colonies had fought them for independence, and he had fought them in the War of 1812. Instead, he pronounced that Britain was "alike distinguished in peace and war."[18] Further, he noted that both countries were deserving of mutual respect. During his administration, the United States and Britain did improve trade, specifically between the Empire's Caribbean and North American colonies. Jackson also pursued pragmatic cooperative projects with Britain such as the construction of lighthouses in the Bahamas and along the coast of Florida.

Perhaps for Jackson it was simpler to assess Britain's interests and develop a policy toward it. Britain did not suffer from the same revolutions, civil wars, and challenges of succession that befell other European nations. Although Britain alternatively had the Tories, Whigs, and Conservatives in power in the 1830s and had two different monarchs, the system of government was essentially unchanged in relation to American foreign policy needs. Consequently, Britain's stability and reliability meant that Jackson could turn his attention to other places, such as France.

Jackson's first public pronouncements lauded France as an ancient ally with "a powerful, intelligent, and magnanimous people."[19] He intimated that France's former government's position led to "unpleasant discussions and possible collision" between the two countries. At issue was the payment of reparations for US property destroyed during the Napoleonic period (fig. 5.1). As a result of the July Revolution of 1830, France's government again changed hands from the Bourbon (King Charles X) to the House d'Orleans (Louis-Philippe). The terms of a treaty signed between the two governments in 1831 remained unfulfilled by 1833. According to the terms of the treaty, France was to pay for depredations on American shipping during the Napoleonic Wars.[20] "Near a quarter of a century has been wasted in effectual negotiations to secure it," Jackson said.[21] In his most extensive treatment of a diplomatic topic, Jackson also recognized the danger of escalating the precarious state of affairs in US-French relations. Elements in Congress seeking to force France's hand considered a trade embargo. Jackson advised a more restrained approach. Cutting off trade with France likely would have meant retaliatory measures on US exports.

Few Americans articulated the demand for forced justice better than one of the era's greatest literary figures, James Fenimore Cooper. Cooper had been a midshipman in the Navy prior to the War of 1812. Thus, he had the

5.1. *The Debilitated Situation of a Monarchical Government . . . The Crisis between France and the U.S.* This print celebrates President Andrew Jackson's successful receipt of reparations from the French for damages to the US Navy dating back to the Napoleonic Wars. Library of Congress, Prints and Photographs Division.

ear of many senior Navy officers who had once served with him. His novels had made him a well-known figure in the country, although from 1826 to 1833, he wrote them from Paris. When he returned to the United States, he warned that a crisis between the two countries had arrived. "It is admitted on all hands," he wrote, "that America and France stand towards each other, at this moment, in warlike attitudes."[22] War with France would have to be fought vigorously on the offensive, that offense would be maritime in nature, and the Navy would have to seek a victory like Trafalgar. That Cooper would suggest the United States had a force of sufficient size to merit such a woefully unrealistic result underscores what Jackson faced in his opponents on this issue. Nevertheless, Cooper had the power of the pen: "I know that the public mind is not yet prepared for a great demonstration of naval force; that opinion has not kept pace with facts. . . . It is my aim to prove their error."[23] Cooper was prodding the country into a war it neither wanted nor was prepared to fight. Jackson's evaluation of the consequences of taking economic or military action against France was sophisticated and measured, particularly given his lack of foreign policy experience prior to becoming president. He believed that the

United States was in the right on demanding reparations. But if any actions injured France, as some in Congress or supporters like Cooper desired, international opinion would change. The United States would no longer have an unblemished, righteous cause. Each injurious action the country might take would cloud and confuse its intentions on the world stage:

> Besides, by every measure adopted by the Government of the United States with the view of injuring France the clear perception of right which will induce our own people and the rulers and people of all other nations, even of France herself, to pronounce our quarrel just will be obscured and the support rendered to us in a final resort to more decisive measures will be more limited and equivocal. There is but one point of controversy, and upon that the whole civilized world must pronounce France to be in the wrong. We insist that she shall pay us a sum of money which she has acknowledged to be due.[24]

Jackson's long-term approach to the crisis with France had abated the difficulties by 1836. Diplomatic relations between the two countries resumed and the two countries reached an agreement on reparations.

The War for Greek Independence entered the halls of Congress in the 1820s. In 1824, Senator Daniel Webster joined Senator Henry Clay and President Monroe in supporting Greek independence from Turkey. Supporting the Sublime Porte were American merchants seeking new trade routes and "northerners rebelling against the National Republican ascendancy and the representatives of the increasingly conservative South led by John Randolph."[25] The decisive naval Battle of Navarino between the great powers of Europe and the Ottoman Empire ended in the former's favor in 1827, but the land war continued into Jackson's presidency.

The Greek War for Independence had a secondary effect on American commerce. In the wake of no stable power to provide security on land and at sea, Greek pirates took to the Aegean and Adriatic to attack shipping from all nationalities. Despite the comparative pacification of Algiers, Tunis, and Tripoli, depredations on American commerce continued in the 1820s but now with Greek pirates. Branch's predecessor, Samuel Southard, reported to Congress that allied nations had diminished piracy largely through a convoy system. Some islands were still a problem with their lawless inhabitants in the wake of the Battle of Navarino.[26]

The Committee on Naval Affairs believed the current squadron had to be sufficient to convoy short distances given the financial circumstances of the government.[27] Southard replied that "there is, unquestionably, at this time, much danger to our commerce from piratical depredations. The continuance of the danger will depend very much on the course of the allied squadrons, and the diligence and energy of our own vessels."[28]

Though Greek piracy had minimal impact on American commerce compared to the actions of pirates in the West Indies, the threat served an important purpose for the Mediterranean Squadron. Just as the First Barbary War enabled the young Navy to coordinate with a European power (Sweden), the Greek threat enabled the United States to establish itself as a Navy among relative equals in the Mediterranean. The squadron commanders communicated with one another. For example, Commodore Daniel Patterson's flagship was in Smyrna only two weeks after the Battle of Navarino. The British provided a plan for the position of the ships during the battle, while the French squadron's commander at the battle—Rear Admiral Henri de Rigny—discussed with Patterson operations against the pirates.[29] The dominant British and French might not have shared this information if they did not consider the US Navy a new ally.

Some 130 Turkish ships of all types, including troop transports, began the battle; only 16 survived.[30] The defeated Ottoman Empire needed a friend, and it found one in the young American republic. The Ottoman Empire needed to rebuild its fleet, and on September 12, 1829, the secretary of state sent merchant Charles Rhind; US commercial agent to Turkey David Offley; and commodore of the Mediterranean Squadron, James Biddle, to negotiate a treaty. The result was the Ottoman-American Treaty of Commerce and Navigation concluded in May. Rhind included a "secret article concerning naval assistance," but Biddle and Offley opposed such a commitment.[31] The Senate, however, refused to approve a provision for direct naval assistance. A commercial treaty with the Ottoman Empire was important for access to wheat via the Black Sea and as a source of Turkish opium to sell in China.[32]

In 1830, shipbuilder Henry Eckford designed and built a new 26-gun corvette on speculation. The US government did not purchase it so, with references from Jackson, Eckford sailed it to the Ottoman Empire where the sultan purchased it.[33] The sultan was so impressed—and in so great need—that Eckford found himself hired as the Ottoman Empire's chief naval constructor. When the sultan first became aware of Eckford's skills, he was reported by the American ambassador to have said, "America must be a great nation if she could spare from her service such a man."[34] Eckford died in 1832, leaving his duties to his protégé, Foster Rhodes.[35]

Turkey rewarded the Jackson administration for its support. By the end of his second year as president, Jackson had put forth a commercial treaty providing the United States equal access to European nations in trading in the Black Sea. Jackson was able to tell Congress that the treaty had had its effect in opening new markets for American commodities. In 1834, Jackson stated that American "relations with the Sublime Porte promise to be useful to our commerce and satisfactory in every respect" to the US government.[36]

Other European powers, such as Portugal, also presented challenges to Jackson. The country was in a state of civil war from 1828 to 1834 between the competing Liberals and Miguelites.[37] The Portuguese Miguelites (those who supported Prince Miguel as the legitimate heir of King John VI) blockaded the Liberal stronghold of Terceira Island in the Azores in 1830. A year later, the Portuguese colony of Brazil faced its own crisis of authority when King Pedro abdicated and went to Terceira. American commerce in the region was also attacked as a result of the hostilities. The United States demanded reparations for damage occurring to American merchants, while Jackson announced that he was preparing to dispatch a force to prevent further violence against Americans and ensure they could continue lawful commerce.[38] With several European powers participating in the blockade, the Portuguese government did not need another naval force in the region. Portuguese representatives convinced Jackson they would honor reparations. Consequently, he canceled his plans to send a squadron.

Jackson clearly understood that America was largely protected from European intervention due to the Atlantic Ocean. Still, he understood that he should respect the size and strength of those empires. In diplomatic overtures, he showered the European powers with public praise. For example, he called Russia "a great power in the rank of nations."[39] He also knew that by endearing himself to these countries he stood a greater chance for beneficial agreements. By the end of his first term, a treaty of navigation and commerce had strengthened American relations with Russia. Such was the case with other countries, like Austria and the Kingdom of Two Sicilies. Jackson's successful expansion of trade agreements with European powers demonstrates a sophisticated understanding of international relations. It also demonstrates that he clearly understood the Navy's role in conducting diplomatic negotiations and protecting that trade.[40]

After the passage of the 1808 congressional act prohibiting the importation of slaves from Africa, traffickers emerged to circumvent the law. The country either could not or would not enforce the new act. The Navy was unable to respond immediately as most of its larger ships had been dry-docked after the First Barbary War. Its growing fleet during the War of 1812 was engaged with British warships. The postwar Navy had some initial success in operating against slavers, particularly those entering Florida and generally the Gulf of Mexico, where slave traffickers were rampant. In 1819, Congress authorized the Navy to operate off the coast of West Africa. The Navy was charged with showing the flag to protect the fledgling colony for freed slaves in Liberia, where a settlement was established in 1821. Slaves had been imported for nearly two centuries, and yet, no records of the slaves' birth areas existed. With little understanding of the locality in which the slaves originated, the United States founded a colony in Liberia named after Monroe, like Sierra Leone founded by the British in 1787 for the same purpose. Since 1819, approximately 260 freed slaves had landed there at the expense to the US government of $250,000. Monrovia ultimately became the destination for thousands of freed slaves and a destination for Navy ships until the Civil War. Officers like Matthew Perry, who commanded *Shark* off West Africa in the 1820s, used the experience to improve health conditions for his and later crews.[41]

During Jackson's administration, the Navy barely made annual visits. When the frigate *Java* arrived in 1831, Captain Edmund P. Kennedy found the colony in a flourishing and happy state. In a letter to the secretary of the Navy, he optimistically assessed that he had no doubt that it was the foundation of a great empire. He also noted the schooner guarding the establishment was too small and recommended a ship of at least eighty to ninety tons with thirty men. The *Boxer*, under Lieutenant Benjamin Page, visited Monrovia in 1832 followed by the *Porpoise* and *John Adams* and the *Ontario* and *Erie* in 1835. In Jackson's last year in office, four Navy ships arrived on station.[42] The Navy provided a physician to liberated slaves. Equipment and supplies, however, were not provided, but, when the *Boxer* arrived in 1832, Lieutenant Page was under the impression from the secretary of the Navy's orders that the ship would provide the freed slaves musket cartridges, priming powder, slow match rope, wine, and cannon powder. Secretary Woodbury apparently expected compensation for the supplies, but the *Boxer* had already left the area.[43]

Jackson was in Nashville in 1816 when the American Colonization Society was founded on December 28 and his involvement was merely ceremonial despite his having been made vice president of the society on January 1, 1817.[44] Jackson reported to Congress in 1831 on the capture of the Spanish slave ship *Fenix*. Jackson recommended that "suitable legislative provision be made . . . for the unfortunate captives."[45] The 110-ton brig was built in Maine in 1802 as the *Antelope*. Sold to foreign interests in 1809, it served as a slave ship for the Spanish government in Cuba a decade later. American-built ships—particularly the later Baltimore clippers—were generally faster than their European counterparts or Navy ships and served the illicit trafficking trade extremely well.

On June 5, 1830, the *Grampus* under the command of Lieutenant Isaac Mayo (who two decades later commanded the Africa Squadron) encountered the US merchant ship *Kremlin* near Santo Domingo in the Caribbean. A piratical vessel, determined to be the *Fenix*, had approached and attacked the *Kremlin*. *Grampus* gave chase to the *Fenix*. Upon approaching it, the ship failed to comply with his order to lower sails. Mayo ordered a shot fired close to *Fenix*'s hull, successfully stopping the ship. The ship was carrying eighty-two slaves. In his letter to Secretary Branch, Mayo suggested the ship had also been involved in piratical acts and that it was his duty under law to bring to trial slavers. Mayo had the captured *Fenix* brought into the port of New Orleans. Attorney General John Berrien determined that the *Fenix* did not fall under the act of 1819 as it was a Spanish ship not captured within US territorial waters nor were the slaves destined for the United States. Berrien did inform Branch that the *Fenix* was also held for piracy, and, as such, the owners had to forfeit any property (including slaves). "The Africans," he wrote, "must be disposed of in some manner which will be consistent with their rights and with the principles of humanity."[46]

Berrien's statement is curious given his background. He was a slave owner from Georgia; at one point, he owned more than one hundred slaves. But more germane to the issue was his experience with the *Fenix* when it was still the *Antelope*. As previously noted, the *Antelope* had a history as a slave-trading ship. In 1820, it was stopped off the coast of Florida with 240 slaves. The eventual Supreme Court case had proved precedent setting in the United States on the legitimacy of the international slave trade. Berrien argued on behalf of the government to maintain the law at that time. The ruling ordered half the slaves to be liberated and sent to Monrovia. Serving on the Supreme Court was Bushrod Washington, a fellow founding member of the American Colonization Society.

The minimal naval activity off Africa might be considered atypical for Jackson's squadrons. Most squadrons were far more robust, the ships remained on station for years instead of a few months, and they conducted diplomatic negotiations with the local governments. That, however, was not possible in Africa since it was largely devoid of nation-states and its tribal organization was often subsumed or overshadowed by European colonial powers. Lacking both governments with whom to negotiate and markets for American goods, western Africa held little interest for the commercially focused Jackson.

The Pacific Squadron's mission was far different from that of the Mediterranean Squadron but no less important. The squadron, which had been established in 1821, was comprised of only two ships—the ship of the line *Franklin* and the sloop *Dolphin*, both under the overall command of Commodore Charles Stewart. At the time South America was in the throes of revolution as figures such as José de San Martín, Simón Bolívar, Bernardo O'Higgins, and José de Sucre sought independence from the Spanish royalist government. Stewart's orders, like those of his successors, were clear: maintain strict neutrality.

The American government's interest in the region was entirely about the protection and continuance of its growing commerce in the Pacific, specifically its whaling industry. American ships, especially its warships, could not have operated on extended cruises in the region without the support from foreign ports. *Franklin*, for example, was on station for two years. Consequently, interference in a conflict would have diminished the squadron's ability to resupply or make repairs.

By 1829 most of the revolutions had passed and a war between Peru and Chile would conclude the following year. The situation was such that Secretary Branch stated: "Peace has generally prevailed among the nations on the western coast of South America. Should this state of things continue, our vessels will have an opportunity to extend their cruises to those portions of the portions of the Pacific most occupied by our merchant ships, and be useful to them in their pursuits."[47]

Still, this did not mean American merchant ships were immune from being victims of conflict. In one case, Peruvian authorities had forcibly taken the American bark *Peru* to transport troops from Quayaquil to Payta.[48] The Pacific Squadron in 1828 consisted of the flagship frigate USS *Brandywine* (44 guns), sloop USS *Vincennes* (18 guns), and sloop USS *Dolphin* (12 guns). The following year the frigate *Guerriere* (44 guns) and sloop *St. Louis* (18 guns) relieved the *Brandywine* and *Vincennes*. The second Adams administration ordered the *Vincennes*, under Captain William Bolton Finch, to proceed to the Sandwich Islands and return via the Cape of Good Hope.[49] Should it complete its assignment, it would become the first US warship to circumnavigate the globe.

It was unusual for an administration to send a ship on such a mission during a lame duck session. On January 15, 1829, Southard wrote to Adams that he was about to issue orders for the *Vincennes*. Southard argued that it was extremely desirable that Finch present gifts to the local chiefs on behalf of the government as a sign of friendship. Southard hoped for preferential treatment for US whaling and merchant ships that visited the Sandwich Islands. Such an approach, he noted, had been successful for the Russians and British. Southard proposed a pair of globes and map of the United States for the king, a silver bowl for the queen, and a silver goblet for the princess.[50] This proposal came fully two months after Adams had lost the national election, a month after the electoral college had met to vote on the president, and seven weeks before Southard and Adams would both leave their positions to the new administration. This decision speaks to the needs of the country rather than the individual. Despite their political differences—indeed the animosity between the two—the Adams administration hoped the Jackson administration would continue policies necessary for American advancement in the world.

The cruise of the *Vincennes* was significant for the Jackson administration for three reasons. First, it is important to understand that Jackson could have rescinded the ship's orders originally given by Adams. There was a distance problem with regard to timely communication with the ship. *Vincennes* was in port Callao, Peru, on July 17, 1829, when it got underway for the Sandwich Islands.[51] New orders could have reached *Vincennes* there with the squadron or through other ships at the Sandwich Islands. An antinavalist would have recalled the ship. That Jackson did not suggests a continuation of at least some of Adams's vision for the Navy, which was unusual since Jackson had repudiated much of his political nemesis's policies. Second, the *Vincennes* was to visit ports that could be valuable to American commerce in the Pacific. The Sandwich Islands offered a strategic commercial outpost for the American whaling fleet as well as warships that sought to proceed westward. The islands were critical for squadrons or fleets of merchant ships, since they could provide respite for crew health and were integral to resupplying ships. Good relations with the inhabitants were important to achieving those needs. A US consul was present on the Sandwich Islands and requested a sloop to be permanently based there "to regulate all things, and secure our commerce."[52]

Third, it demonstrated the unstated intent of the Jackson administration for national growth. As a general, Jackson had helped the United States push south to Florida and the expectation was that the nation—prior to a formal announcement of Manifest Destiny—would continue to advance to the western coast of the continent. *Vincennes*, and later other ships following it, was meant to broadcast that the Pacific Ocean was easily traversable by the US Navy, to open important markets with the Far East, and to demonstrate that a Navy ship could operate globally.

Lieutenant Thomas Dornin, one of *Vincenne*'s officers, maintained the best existing accounts of the journey.[53] In addition to Master Commandant Finch, the wardroom included at least three future key officers. First Lieutenant Cornelius Stribling's experiences would help him to command his own ship to the Far East with the creation of the East India Squadron in 1835. Midshipman Matthew F. Maury would become one of the most prolific writers among the junior officers that decade and one of the most influential officers of the Navy in the mid-nineteenth century, although he would turn his back on the Navy by joining the Confederacy. Thomas Melville would recount his experiences to his brother, noted author Herman Melville, who based a number of his books on actual naval or merchantmen accounts including *Moby-Dick* and *Typee*.[54]

While in Valparaiso, Chile, the Vincennes welcomed *Guerriere* and *St. Louis* on June 19, 1829. Finch read the orders to sail to the Sandwich Islands, then to Manila in the Philippines, the Cape of Good Hope, and westward to the United States. "We all anticipated much pleasure and information from our intended cruise," Dornin wrote.[55] Also joining the ship as a supernumerary was Reverend Charles S. Stewart (no relation to Commodore Stewart) on his way as a missionary to the Sandwich Islands. Like Maury, Stewart would become one of the most well-known writers of the decade and later serve as editor of the *Naval Magazine*. According to Dornin, Stewart was "well known as the indefatigable missionary of the South Sea Islands."[56]

In late July the ship pulled in to the Sandwich Islands. Finch and several of his officers paid their respects to the uncle and guardian to the infant King Kamehameha III. Intertribal rivalry was still prevalent, and one tribe hoped the United States would aid them against their alleged warlike neighbors. Upon viewing a French-caused incident, Dornin reflected, "It is to be lamented that foreigners should be so thoughtless as to supply these people so abundantly with muskets and powder which stimulates them to be always in open war with themselves and extremely dangerous to the Merchant ships that visit them."[57]

On September 1, the *Vincennes* reached Tahiti. Its king told Finch that he was always pleased with the arrival of merchant ships. More importantly, they had never had any complaints or incidents with regard to American seamen. By January, the ship was off Macao in the Far East. Dornin noted that the Portuguese authority was nominal with a mandarin in real control over the colony. Throughout that winter, the ship traveled to Canton and Manila. This was not the first Navy ship to visit the region. The frigate *Congress* under Captain John D. Henley arrived at Lintin Island in 1819 for a visit that was not entirely successful.[58] American businessmen viewed it with suspicion as a threat to the existing order; China rejected foreign warships; and they viewed foreigners as *fan kwei* (foreign devils).[59]

When *Vincennes* made it back to the United States, Branch commended the ship and her captain for having "returned in good condition, with its crew well-disciplined and in excellent health."[60] Finch had succeeded in his mission in securing the continuance of kindly treatment of his countrymen. The peaceful cruise of the *Vincennes*, however, was not echoed by the mission of the frigate *Potomac*. On February 7, 1831, the Salem-based merchant ship *Friendship* under Captain Charles Endicott had anchored off Quallah Battoo on the west coast of Sumatra's Aceh Province. During Jackson's presidency, few American cities were as prosperous as Salem, Massachusetts, or as maritime oriented on an international market. Jackson's stated goals relied heavily on maritime cities like Salem. The city built the subscription frigate *Essex* during the Quasi-War and presented it to the Navy.[61] Its visits plied the Far East trade that included opium and valuable pepper. The city also gave birth to merchants like Elias Hasket Derby, the first American millionaire and one of the first to trade with China.

The *Friendship* had a crew of seventeen. Endicott went ashore with two officers and four crewmen to negotiate with the local raja for pepper. By the end of the day, Endicott and the raja agreed to a price and local natives began to take the pepper back to the ship. Although instructed not to come aboard, the natives boarded the *Friendship* and killed three crewmembers, including the first mate, Charles Knight. The pirates quickly but thoroughly absconded with the ship's treasure. The financial damage to Endicott was just as costly—$12,536 in specie, $8,818 in opium, $2,500 in stores and provisions, and $17,000 in other losses.[62]

Endicott and his men made for their small boat before the natives on land could take them. After being chased by three small local boats through the harbor, Endicott made for the trading village of Muckie, twenty-five miles away, where they found three American merchant ships the following day. The ships made for Quallah Battoo and after a battle retook the *Friendship*.[63] Endicott retook the ship and returned to Salem empty-handed on July 16, 1831.

On June 27, 1831, Woodbury issued orders to Captain John Downes in Norfolk to proceed to New York to finish equipping the frigate USS *Potomac*.[64] Woodbury's instructions to Downes were lengthy given the transit time to his eventual station and the circumstances that might develop, but there was no mention of the *Friendship* attack, as news had not yet reached the United States. The ship was to take aboard Martin Van Buren as minister to Britain before its eventual assignment in the Pacific. There, Downes would relieve Francis Gregory in command of that station, and Van Buren would join Gregory's ship *Falmouth* and Lieutenant John Long's schooner *Dolphin*.[65] That force represented nearly 10 percent of the Navy. Woodbury had determined to provide sufficient protection to US interests in the region.

To prepare Downes for his duties on station, Woodbury provided him with a volume on the treaties between the US and foreign powers and a compilation of Navy laws. Other volumes included regulations on seamen in foreign ports, how to make quarterly reports on boarded vessels, and guidance for ship captains with consuls. "Cases may arise," Woodbury wrote, "which it is impossible to foresee, and to meet which definite instructions cannot be given; should such occur . . . you must be left to the exercise of a sound discretion."[66] This directive is yet another example of Woodbury's understanding of the need to have well-informed officers who would have no immediate communication with their chain of command once they arrived on station.

Jackson learned of the incident at Quallah Battoo on July 19 not through official channels but through newspaper reports. The merchants of Salem appealed to Jackson to demand retribution. Jackson immediately issued orders for the *Potomac* to seek redress for the outrage committed.[67] Woodbury reissued orders to the *Potomac* on August 9, when the *Potomac* reached New York. Gone was the general guidance and anticipation of a leisurely circumnavigation of the world. Expediency was now a priority. Woodbury directed Downes to proceed immediately by way of the Cape of Good Hope, going ashore only as necessitated by low supplies or other immediate circumstances. As conveyed through Woodbury, Jackson ordered Downes to demand restitution from the raja or other authorities at Quallah Battoo and immediately punish anyone involved in the murder of First Mate Charles Knight and the two other crewman of the *Friendship*. Also included was direction to destroy boats of any kind engaged in local piracy and forts and dwellings near the incident used for shelter or defense. Finally, Downes received orders to inform the locals that, if restitution would not be made, a greater American naval force would return "to inflict more ample punishment."[68]

Though this was a direct military response to an attack on a US merchant ship, Jackson was also aware of the longer-term implications, notably a healthy commercial relationship with the native population. Jackson wanted it conveyed to the raja and local population that there were no hostile feelings toward the people of Quallah Battoo—only the plunderers and murderers. Woodbury wrote to Downes that he must take great care with caution, forbearance, and good faith toward the natives.

Downes, a thirty-year veteran of the Navy, had fought in the Pacific during the War of 1812 on David Porter's *Essex*. He was, thus, prepared nearly as well as any other captain for this mission.[69] After a lengthy voyage, *Potomac* arrived in Quallah Battoo on February 5, 1832, nearly a year to the day after the attack. Downes anchored off the shoreline some three miles to prevent detection as a ship of war; he had also taken other precautions to disguise the ship. Downes exceeded his orders.[70] Despite being told to seek retribution, Downes attacked the town since he did not believe the locals would agree to any demand.

Lieutenant William Shubrick, the *Potomac*'s forty-two-year-old first officer, took a contingent of midshipmen and Marines ashore to assess the situation and returned to the ship. The following day Shubrick commanded 250 men aboard all the boats and rowed a mile and a half north of the town so as to meet with the least resistance. The force captured the northernmost fort after a two-hour battle. Two more forts awaited the raiding party, but they pushed on to the strongest fort at the southern end of the town. One sailor was killed and four more wounded in the attack.

Shubrick directed the fire from his mortar at the buildings close to the fort resulting in two magazines exploding. Downes moved close enough toward the shore to fire three broadsides into the fort. After the third broadside, the white flag rose in surrender. The American flag quickly replaced it as Shubrick's force entered the fort and spiked the guns. The final fort fell after another two and a half hours of fighting with one Marine killed.

Shubrick noted in his report to Downes that the action had reduced the town to ashes. At least 150 of the enemy died including Po Mahomet, the primary culprit for plundering the *Friendship*. The attackers took no prisoners, but neither did they recover any property from the *Friendship* except for its medicine chest. Downes reported to Woodbury that he negotiated with a local leader for peace. Several rajas visited the ship, and all agreed to be friendly with the Americans.

Potomac continued its mission to assume duties on the Pacific Station off the west coast of South America. When the ship returned to the United States, it had completed the second American circumnavigation of the world. Downes faced public criticism for his action in failing to exercise judicious response and exceeding his orders. Nevertheless, Jackson largely supported him since there had been no further attacks on American ships.[71] On December 4, 1832, Jackson delivered his fourth annual message to Congress and briefly mentioned the Sumatran expedition. The United States had met the original atrocity with an appropriate response to deter like-minded pirates. Jackson commended the operation arguing that it had the effect of an increased respect for the American flag in providing security for commerce in those distant seas.[72]

Two other lesser-known incidents occurred during Jackson's second term. The Nantucket whaler *William Penn* anchored off the island of Savai'i in the Samoan archipelago in 1834.[73] The locals captured and killed the first and second mates, but some of the crew survived. Another whaler saved them and took them to Tonga where the *Vincennes* was briefly visiting. Upon hearing of the attack, Captain John Aulick immediately sailed to the Samoan island. Basing his judgment on the precedent set by Downes, Aulick sent eighty men ashore. The villagers had already left, but the sailors, on order from Aulick, burned the village.[74]

The second incident occurred in October 1835 when the New Bedford whaler *Awashonks* pulled in to Namorik Atoll in the Marshall Islands.[75] Local villagers boarded the ship, attacked the crew, and killed five men, including the captain and two officers. The remaining crew reached the armory and were able to fend off the villagers. The Navy sent no ship on a punitive raid. Based on conventional assessments of Jackson's temperament, he should have sent a warship to Namorik. In the cases of the Falklands Crisis and Quallah Battoo, one Jackson biographer stated: "This is Andrew Jackson acting as Andrew Jackson. He sees an offense against his country and he's the president, commander in chief of the Army and Navy and he takes action."[76] But Jackson issued no similar order for the Savai'i or Namorik incidents.

Jackson did not indiscriminately use force with regard to the Navy. He weighed his options and considered the consequences within the framework of his maritime strategy to expand trade and protect American commerce. Namorik was a very minor atoll in the middle of the Pacific where a few ships might stop to take on water but then only infrequently. There were not enough villagers with whom to trade and no local resources that US merchants desired. In other words, there was no commerce of any consequence. An attack on the island would have meant little except to the local villagers. It did not rise to the level of strategic economic trade as Sumatra or even the Falkands. It is doubtful that Jackson would have even issued a punitive raid on Savai'i had not Aulick done it first.

Even though Namorik was comparatively insignificant to the United States, it and the hundreds of other Pacific islands and atolls needed to be surveyed, should any be of use for the servicing of merchant ships during their lengthy cruises. Newspaper editor Jeremiah Reynolds first proposed such a scientific expedition in the 1820s (Reynolds later served as Captain Downes's personal secretary aboard *Potomac* during the Sumatra expedition). Its objective would be to improve American commercial interests. The Navy would chart, survey, and explore the greater Pacific region. "[Reynolds] knew that the support of merchants and ship owners would greatly improve his prospects for winning government funding."[77]

Just two weeks before Jackson was inaugurated, the Senate Committee on Naval Affairs considered legislation authorizing such an expedition. The Senate, controlled by Jacksonians, demurred when it learned that the Adams administration had made it appear the expedition was nearly ready to depart. In fact, no details had emerged about the makeup, provisioning, or cost of the expedition. Senator Robert Hayne made it clear in the committee report that he and his colleagues were supportive of the idea of an expedition, but, without more knowledge about how it might work, they could not endorse the legislation.[78] After directing the secretary of the Navy to provide information, the committee learned the expedition would cost approximately $438,000, or one-seventh of the total naval budget for 1829. The Senate, therefore, killed the bill.

The project was cost prohibitive to an incoming administration that intended to scrutinize the Navy's budget. By 1834, the administration reduced the federal debt with the expectation that a surplus in 1835 would eliminate it. With a surplus, the administration could revisit projects like the exploring expedition. Merchants began appealing to Congress. In 1834, the East Indian Marine Society of Salem wrote that it was praying for an expedition. It called maritime commerce the "fountain from which unfailing streams of revenue our financial reservoir is supplied with the means of national existence."[79] Groups of citizens and merchants from other, mostly northern, cities joined the cause. New York argued the case "for the benefit of navigation, whale and seal fishery, commercial trade [and] science."[80] Citizens of Philadelphia echoed the appeal for navigation and exploration but also for "commercial enterprize."[81]

It would be two more years until Congress took up the Senate's legislation approving the South Seas Exploring Expedition.[82] By a vote of ninety-two to sixty-eight, the House authorized the project and appropriated an additional $150,000 for fitting out a ship.[83] Jackson signed the long-awaited project on May 18, 1836. The following month, he and Secretary Dickerson began interviewing possible commanders. Jackson's first choice was Captain Thomas ap Catesby Jones, whom Jackson had met after the Battle of New Orleans. Jones eventually declined due to poor health. Four other experienced captains also declined. Command of the 1838 to 1842 expedition eventually fell to Charles Wilkes, a junior lieutenant.[84]

The Brazil Station consisted of the frigate *Macedonian* (36 guns) and sloop *Boston* (18 guns), which were relieved in 1829 by the newly constructed frigate *Hudson* (44 guns) and sloop *Vandalia* (18 guns). Commodore Stephen Cassin relieved Commodore John Order Creighton in 1829. Jackson increased the ships on station to respond to the threat to American commerce. A war between Brazil and Argentina had resulted in blockades by those nations. Even when the war ended in 1830, Jackson felt the nation needed a larger US force because of increasing commercial relations with countries in South America. He, therefore, increased the number of ships. It was fortuitous that he did so given a new crisis that emerged in the South Atlantic.

Since the European discovery of the New World, different nations had laid claim to the Falkland Islands. But in 1831 the claim rested with the United Provinces of the River Plate (a predecessor of Argentina that included parts of modern Chile, Uruguay, and other states), whose capital was Buenos Ayres. The commercial value of the Falkland Islands lay in its fishing industry and sealing. Buenos Ayres authorized the establishment of a colony at Puerto Soledad in 1828 under Luis Vernet, the military and civil commander for the Falkland Island. The British government opposed this move because it had its own claims on the territory. Buenos Ayres also expected its privateers to operate out of the Falklands in their conflict with Brazil. Vernet's deputy was the British-born Matthew Brisbane, an experienced explorer in the region.[85]

Attacks on American merchant ships from that region were familiar to Jackson. One example occurred just two weeks before he assumed the presidency. The *Federal* under the flag of Buenos Ayres, allegedly a privateer, attacked the American merchant vessel *Nymph*. The *Erie*, under Captain Daniel Turner, captured the *Federal* off St. Bartholomew Island on January 5, 1829, and took it to Pensacola.[86]

Jackson had been advised of the United Provinces territorial claims as early as January 1831.[87] In June, Woodbury ordered the 24-gun sloop *Lexington* to the Brazil Station to relieve *Vandalia*. Built in 1825, *Lexington* was comparatively new. With only one deployment to the Mediterranean, it was decommissioned in 1830 and recommissioned in May 1831, just two months before it deployed. Captain Silas Duncan, a veteran of the War of 1812 who served under Thomas Macdonough during the Battle of Lake Champlain, commanded *Lexington*.

Prior to deployment, Duncan objected to the assignment of a Marine officer aboard since sloops only rated a sergeant overseeing the Marines. The only person who could have contradicted the secretary of the Navy on this issue was the president. As one historian suggests: "It seems plausible, if not likely, that (1) Jackson personally decided to place a Marine officer aboard the

Lexington, and (2) he did so as a precaution against the possibility that a Marine landing on the Falkland Islands might prove necessary in order to rescue Americans being held captive there. If true, Jackson's role in the dispatch of the *Lexington* was much greater than previously assumed."[88]

The Connecticut-based *Harriet*, under Captain Gilbert Davison, had arrived in the Falklands in November for sealing and whaling. Vernet "warned Davison not to take seals in any of the territory under his jurisdiction."[89] Davison was doubtful of Vernet's authority given American ships had operated in the region for a century, there was no recognized claim by the United Provinces, and the German-born Vernet spoke with a heavy accent. The *Harriet* continued its operations and returned in February and again encountered Vernet. When *Harriet* again returned on July 13, 1831, a small armed force led by Brisbane approached Davison's crew. Vernet informed Davison that his operations were illegal, and he had the ship seized for adjudication in Buenos Ayres.

When a second American whaler—the *Breakwater*—arrived on August 17, 1831, the captain and several crew also found themselves under arrest. The first mate managed to retake the ship and return to Stonington, Connecticut, and report to its owners. On August 19, a third American ship—the *Superior* under Captain Stephen Congar—was seized and its crew, like the *Harriet*'s, was imprisoned. Vernet made an agreement with Congar to continue sealing if half of the skins went to Vernet. Based on Vernet's assessment, foreign whalers and sealers were depleting local resources. Arguably, Vernet sought a monopoly on the trade for the United Provinces. Vernet took the *Harriet* to Buenos Ayres and arrived on November 20.

On December 6, Jackson delivered his third annual message to Congress and noted the tense situation that awaited the Navy. The government in Buenos Ayres threatened legitimate, long-term commerce in the region. "I submit the case to consideration of Congress," he wrote, "to the end that they may clothe the Executive with such authority and means as they may deem necessary."[90] He had already ordered a Navy ship to the region to protect US interests. But, as Klafter observes, it was not the *Lexington* that had been ordered to the Falkland Islands but the sloop USS *Enterprise* under Commander George W. Rodgers of the Perry-Rodgers line of naval officers.

Secretary of the Navy Woodbury ordered *Enterprise* to Brazil Station and then to proceed alone or with *Lexington* and *Warren* to provide "complete protection" and to coordinate with local American officials.[91] When *Lexington* arrived in Buenos Ayres on November 27, Captain Duncan received news that American ships had been seized. In addition, he saw *Harriet* lying in the harbor between two Argentine sloops. To attend to the imprisoned Americans, he made for Puerto Soledad (also called Puerto Louis after Vernet) on December 31, 1831. Duncan arrived flying the French flag to confuse the colony's armed contingent.[92] The following day, he landed the Marines who rescued the American crews. Duncan proceeded to take the town and spiked the cannons. For the next two weeks, the *Lexington* took aboard Americans and settlers and made its way to Montevideo, arriving on February 2, 1832. The local government claimed Duncan had gone well beyond his mission by destroying private property. He arrested Brisbane for piracy and the others who had supported him for attacking a fourth American ship, the *Elizabeth Lane*.

Rodgers arrived to relieve Duncan from his command of the squadron.[93] Rodgers immediately dealt with leaks to Washington, in which officers had reported to the Navy and media on operations. On March 16, he issued the order to the squadron that only the commodore could communicate with the secretary of the Navy. Furthermore, he forbade all officers from communicating to the public through friends or others any movements of the squadrons.

Rodgers was still trying to assess his mission two weeks after arriving. He summarized the situation in a letter to the American vice consul in Buenos Ayres. He recognized the anti-American sentiment in the city, but he believed that the depredations against American commerce around the Falklands had ceased. He did not know if there would be any attempt to recolonize the Falklands.[94] In an act of good faith, he released Brisbane and his compatriots.

The Jackson administration clearly articulated this crisis as a localized threat to legitimate commerce. In no correspondence did he lay the blame on Buenos Ayres, which would have given him cause to blockade their coast and demand reparations. The operation was mature in its assessment of the situation, measured in its response, and effective in resolving the crisis. This became typical of the Jacksonian approach to naval affairs.

The West Indies Squadron was the closest operational squadron to US territorial waters. At the beginning of the Jackson administration, the squadron comprised the 18-gun sloops *Natchez*, *Erie*, *Hornet*, *Falmouth* and the 12-gun schooners *Grampus* and *Shark*. All were ships built for shallow waters. This was a result of the lessons in the previous decade by Commodore David Porter as he chased pirates around the Caribbean. He quickly learned that his ships had too much draft to chase smaller ships close to shore. But if the voyage of the *Vincennes* was an indicator of America's vision for naval operations in the next century, the Gulf of Mexico was the first step to its maritime destiny. In Secretary of the Navy Southard's end-of-year report to

Congress in 1828, he reported on the piracy situation. "No piracies have been committed," he pronounced. "That scourge of our commerce has been entirely repressed."[95] He asserted that the reports of attacks were simply rumors founded on misrepresentation. He was wrong.

Acts of piracy in February 1829 required deployment of the *Natchez* to the West Indies after briefly returning to its homeport. On March 16, Acting Secretary of State James Hamilton, son of Alexander Hamilton, advised Branch that, because of the piratical atrocities, no armed vessel had visited the port of Matanzas since the previous August. More attacks had occurred near Cuba. Hamilton requested additional ships available to counter the threat: "It is due to humanity, the interest and honor of the country, immediately to take the most summary and efficacious means within our power to suppress and prevent the recurrence of these atrocities."[96] The consular office in Matanzas reported that the brig *New Priscilla* of Salem had been captured and the captain and crew of the brig *Attentive* of Boston murdered, among others, such as the schooner *Charles* of Philadelphia, in which the entire crew was also murdered.

Less than two weeks after his inauguration, Jackson requested a status report on the US Navy forces in the West Indies. Rodgers reported to Branch on March 16, 1829, that the current force (*Hornet, Falmouth, Erie, Shark*, and *Grampus*) was sufficient for the suppression of piracy around Cuba. "There are none other of our small vessels," he wrote, "that could be speedily sent out except the *Peacock* at New York and the *Concord* at Portsmouth."[97] *Concord*, however, had never been to sea, and neither was capable of refitting for sea within a month. A day later, Jackson wrote to Branch to order Commodore Ridgely—the commanding officer of the West Indies Squadron—to keep one warship on the coast of Cuba. Pirate activity required vigilance and energy. He directed Commodore Ridgely, commander of the West Indies Squadron, to keep one of his small vessels of war on the coast of Cuba and never to be absent from it longer than a week at a time. "The dictates of humanity and the honor of our flag require," Jackson wrote, "that the piracies in those seas should be suppressed."[98]

The letter conveyed specific orders for named warships demonstrating that Jackson had direct involvement in naval operations. Thus, Jackson's first order to use force as commander in chief was for naval operations not Army movements. If he was not aware of the need for a Navy conducting constant presence before he assumed office, then the pirates threatening US merchant ships to the south made him painfully cognizant of the fact. It was not a lesson he would forget in his administration.

The piracy threat elicited responses from individuals as high as Jackson through the junior officer ranks, although the officers' opinions were not always welcome. For example, Lieutenant William F. Lynch proposed the Navy employ decoy ships to lull unwitting pirates into close quarters. Branch shared the proposal with Rodgers. Rodgers still believed the squadron was "amply sufficient for the suppression of piracy in that quarter."[99] He did admit that a decoy ship was initially adopted at the height of the piracy threat but effected nothing. "Lieut. Lynch could not," he noted, "have been apprised of these circumstances." In an admonishment, Rodgers concluded, "he has manifested more zeal than jugement [*sic*]."[100]

By 1830 Branch realized that nothing short of positive and continued force was necessary in the region. Branch recommended a change in the force composition with the Navy sending three schooners in lieu of a larger sloop. They were lighter in draft. More ships meant patrolling more area. Smaller ships meant access to more harbors during storms. It also meant that the Navy would now have to build them and propose it to Congress. In addition, the threats demonstrated a need for a base in the Gulf of Mexico. The Navy identified Pensacola as a rendezvous harbor, but it required deepening the channel. It was a strategic necessity if Jackson wanted to exert American influence in the region. It represented the ability to address Caribbean piracy as well as the growing uncertainty in the western gulf. On January 23, 1830, Jackson ordered Branch to increase naval presence in the region, particularly because of—as he termed it—the revolutionary spirit there. "I consider we cannot place too soon in those seas the Frigate *Brandywine*, in order to guard our commercial interests, and keep down combinations for piratical objects."[101] Jackson also ordered that the ship's captain proceed to Tampico and Vera Cruz because it was important that the ship be seen in those places.

Sailors risked their lives engaging with pirates in the Caribbean or simply by being aboard a ship like the *Peacock*, whose crewmen were victims of tropical diseases. They were also susceptible to the elements, especially storms. The *Hornet* had been one of the champions of the War of 1812, defeating the HMS *Peacock* during the first engagement of the war. It went on to serve in the Mediterranean and finally throughout the 1820s in the West Indies. One of her senior officers, Lieutenant Daniel H. Mackay, had been assigned to it only a few weeks before it got underway for its latest assignment. "I must begin to pack up and make my little arrangements not I confess without some foreboding, not exactly of death and yellow fever, but it's a climate I do not like."[102] Mackay did not expect assignment to the *Hornet* and wrote to a fellow officer that "Sammy Southard" was supposed to send him to the *Peacock*, believing that the recent election had something to do with the decision. "I was a Jackson man," he wrote. Mackay contended that Southard had had the "arrogance and presumption to take some officers to task for voting for the General." He suggested that two senior officers might have played a role in his orders as a result.[103]

Hornet's commander was Master Commandant Otho Norris. Mackay had strong views about his commanding officer within the first month, stating Norris to be weak and an imbecile: "When you go to sea, if there is choice left you by all means choose the knave and the tyrant, for tyranny if consistent & intelligent may be rendered tolerable but ignorance & imbecility are to my quick and impatient temper utterly insupportable absolute torture."[104]

Hornet was part of the West Indies Squadron patrolling the waters and deterring actions of pirates and was to receive new orders on August 11, 1829. *Hornet*'s wardroom was aware of the situation, believing the Mexican Army was exceedingly anxious for a melee. Spanish forces and Mexican insurgents under General Santa Anna were engaged in military operations near Tampico by late summer. Norris's orders were to evacuate the American consul, his family, and other Americans. The ship anchored offshore when an unexpected gale rose up. Soon the gale had grown into one of the most powerful hurricanes in years. Norris recalled all small boats ferrying people and got underway to escape it but was too late. The *Hornet*, one of the most noted ships in the Navy, was lost with all hands. Ninety days after the ship had sailed from Tampico, there had been no word of survivors.

Congress responded to the loss of the *Hornet* with an act for the relief of the widows and orphans of her crew.[105] The loss did not diminish the naturally contentious disposition of the House of Representatives. Congressman Nathaniel H. Claiborne of Virginia was elected as a Jacksonian in the wave of 1828, though he became an anti-Jacksonian during the president's last term (1835–37). Claiborne opposed the relief because the loss was a result of nature not war. "We have no right," he argued, "to be generous and charitable with the money of other people."[106] Navy and Army officers, he said, were already the wealthiest men in the community.

Congressman Clement Dorsey, reelected in 1828 as an anti-Jacksonian, outlined a history of Navy ships lost to providence rather than battle such as the *Wasp*. Congress compensated widows and orphans of the crew in each case. His conclusion was that "there is nothing in the history of this gallant ship and of her noble crew that can justify a nation in withholding the like expression of national regret for their loss with the same sympathy with the relatives of those who perished with the ship were expressed in the case of other ill-fated vessels."[107]

Congressman Edward Everett of Massachusetts, an anti-Jacksonian, was considered one of the finest orators of the nineteenth century.[108] He dramatically appealed to the House as it considered the vote. "A battle has no terrors to the gallant seaman," Everett said on the floor of the House. He continued: "It is full of hope, promise, glory, and even reward, if no higher motive operated. And cheer him in one of those tremendous tempests to enable him to bear the labor and brief the dangers that surround him? There was not one individual man or officer on board the unfortunate Hornet whose muscles were not stiff with labor whose nerves were not strained with agony before he went down to his watery grave. . . . What if they knew as they were going down the house was debating this provision?"[109] The bill passed 138 to 42.

Jackson had to be concerned with the entire Gulf of Mexico. Mexico had secured its independence from Spain in 1824, but its territory included the province of Texas. The desire by US citizens to settle in Texas and the subsequent move by settlers to secede from Mexico caused a rift between the two countries. Initially, the newly inaugurated Jackson had the benefit of having an American in command of the Mexican Navy. David Porter, a hero of the War of 1812, had been the target of an 1825 court-martial for actions taken during the Foxardo incident in the West Indies. Found guilty, he left the Navy and became the admiral of the Mexican Navy. But it was more than that. Porter offered his services to Mexico if it "would give him a rank sufficiently high to justify the step."[110] In addition to rank (he would technically become the second officer in the American Navy to become an admiral—but in another Navy), he was offered the new Henry Eckford–built brig *Guerrero*.[111] Porter, now in command of a squadron—and a Navy—was at times harassing and evading a Spanish squadron under Admiral Angel Laborde. In 1828, he put his nephew David H. Porter in command of *Guerrero*; although he put up a valiant defense, he was killed and the ship destroyed. One of the survivors was Admiral Porter's son, David Dixon Porter, who would become the second naval officer to become an admiral during the American Civil War. Porter soon outwore his welcome in Mexico and resigned in September 1829. A month later, Spain attacked Tampico without opposition. Porter would find himself the subject of physical attacks by Mexicans on at least two occasions.[112] Porter left Mexico and wrote back to US minister to Mexico Joel Poinsett: "I shall anxiously expect your arrival, and shall pray most fervently for your escape from that den of devils, free from harm."[113]

Jackson faced a greater challenge by Mexico's response to America's diplomat. John Quincy Adams had appointed Poinsett to negotiate for the territories of Texas, New Mexico, and California. Jackson, upon the request of President Guerrero, recalled Poinsett.[114] Mexico outlawed the immigration of American citizens into Texas in 1830, thus fueling dissent not only from those already in Texas but also from those in the United States wishing to emigrate. In addition, the Mexican Navy attacked US merchant ships immediately following Porter's departure and return to the United States. The animosity between the United States and Mexico remained steady throughout Jackson's administration. In 1829, Mexican general Antonio Lopez de Santa Anna defeated a Spanish force of three thousand. By 1832, Texas was requesting statehood, thus fomenting revolution. In 1833 Santa Anna became president of Mexico.[115] Complicating the Texan revolt was the fact that one of its leaders, Sam Houston, had served in Jackson's militia during the War of 1812. During his term, Jackson had the opportunity but not the means to support Texas's secession overtly from Mexico and annexation by the United States during his term. The Army effectively engaged and managed Indians; it was not yet ready, however, to support Texas against the large Mexican force that had defeated the Spanish.

Jackson's own victories had come either from overwhelming Native American tribes or in a well-prepared defensive posture against the British at New Orleans. If he sent an Army through Texas, the Americans would be on the offensive while the Mexicans would be in the more advantageous defensive position. Naval support for land operations would have been difficult at best. Texas had only two small ports in the 1830s—Anahuac and Copano—and neither were connected to the interior of the territory. There were no fully navigable rivers.

Such a venture would have also been contrary to his stated position against inviting aggression against the nation. Finally, and perhaps most importantly, a decisive defeat of the American Army in Mexico would have had a devastating effect on his presidency and himself. Jackson had been propelled into office riding a national, populist wave for the man who had defeated the British invasion in the war. Losing to Mexico would have tarnished his image, perhaps irrevocably, particularly since his real focus was securing the southern territory of Florida.

The Florida territory represented the great unfinished project for Jackson, who had led forces there after the War of 1812 during the First Seminole War. Jackson's obsession with the Seminoles led to one of the longest land wars in America's history, with a 14 percent casualty rate (1,446 total with about 20 percent of those to enemy fire).[116] More than 10,000 soldiers,

Marines, and sailors would serve during the seven-year conflict. Spain had ceded control of the territory of Florida to the United States because of the Adams-Onis Treaty of 1819. The native Seminole Indians had largely moved to a reservation in the middle of Florida. American settlers demanded that the government move the native tribes west. Southern slaves had also been escaping south across the border since the days when Spain was in control, thus giving additional reason for the government to respond to southerners.

In 1830, Congress passed, and Jackson signed into law, the Indian Removal Act. Officially, this authorized Jackson to enter treaties with Florida tribes. According to Jackson, "treaties answer no other purpose than opening an easy door for the Indians to pass through to butcher our citizens."[117] The Seminoles chose to resist. Rather than assume a defensive position, the Seminoles began to attack settlements and forts. On December 28, 1835, Seminoles ambushed and killed more than one hundred soldiers under Major Francis Dade.

Propelled by the "Dade Massacre" and other Seminole attacks, the Jackson administration went to war. Jackson mobilized more troops and ordered additional ships to support the West Indies Squadron. Although this was largely a land war, the Navy and Marines participated in operations and often coordinated with their Army counterparts. The Navy conducted its first riverine operations and patrolled off the coast for weapons smugglers supporting the Seminoles. Naval orders preceded the Dade Massacre. One of the ships from the West Indies Squadron received orders to coordinate with Brigadier General Clinch—then commanding Army forces in Florida. Another ship would be "useful in cruising along the coast from Charlotte Harbor to Tampa Bay."[118] As warships deployed to the region, Dickerson informed Dallas that General Winfield Scott would be the overall commander of troops in Florida with Clinch serving as one of Scott's three subordinate commanders. Dallas was to cooperate with General Scott "in the subjugation and removal of those hostile Indians in such way as you may think most conducive to the public interest."[119]

By January, the West Indies Squadron had been reinforced by the revenue cutters *Washington, Dexter*, and *Jefferson*. This was significant as revenue cutters fell under the authority of the Treasury Department. Secretary of the Treasury—and former secretary of the Navy—Levi Woodbury authorized the transfer of revenue cutters to the Navy.[120] The cutters would supposedly assist with coastal patrols. Eaton had reported to Washington that Spanish fishing vessels had been provisioning Seminoles, though this was never proven.[121] By May, Washington had recalled the revenue cutters. No evidence suggests a reason.

Dallas was commended in January for protecting American settlers in Florida. Dallas asked the Navy for another ship. The *Vandalia* was thus ordered to proceed from Pensacola to Tampa Bay. A week later, the *Concord*, under Master Commandant Mervine Mix, received orders to join operations in Florida as well.[122] By May, the sloop *Boston* had arrived in the region with the promise of another sloop to join Dallas soon. A few months earlier, Dallas recognized that geography was his worst enemy and asked for authorization from the secretary of the Navy to charter additional vessels as he deemed appropriate.[123]

A steamboat was employed in support of military operations once before in Porter's West Indies Squadron.[124] The *Sea Gull* had been purchased by the Navy in late 1822 and arrived under Porter in early 1823. Florida's extensive coastline, numerous inlets and bays, and extensive riverine system were a natural environment for the use of steamships. On May 25, Jackson himself ordered that Dallas detach a large portion of his Marines and seamen against the Seminoles and Creeks using three steamboats.[125] The steamboats deployed to Pensacola. They then operated on the Chattahoochee River, which traversed the Georgia and Alabama border through Florida before emptying out into the Gulf of Mexico, thus providing a key logistics line. The steamboats transported supplies, kept communication lines open, and operated against the native tribes.[126] The steamboats' structures were to be reinforced to defend against enemy weapons and to be mounted with as many artillery pieces as necessary, per Jackson's order. The *Major Dade* joined the steamers *American*, *Southron*, and *Talla Busha* by June. Continuous operations sometimes impacted their ability to remain on station. In October, for example, the *American* broke its shaft and had to return to New Orleans. Nevertheless, it was back on duty less than two months later.[127] A less advanced industrial base might have caused a delay in its return. Therefore, this quick turnaround suggests that by the Second Seminole War the Navy had the capability of making quick repairs.

Small boats from the squadron commanded by junior officers deployed closer to the coastline and up the rivers. Lieutenant Levin Powell of the *Vandalia*, under the command of Thomas T. Webb, took two boats up the Manatee River to search for a Seminole raiding party. With 40 officers and sailors, he engaged with Seminoles and took prisoners. Powell operated against the Seminoles for more than 6 months. In the fall, he returned to Charleston to assemble a larger riverine force—14 boats and 14 canoes, transported by 4 chartered sloops to Florida.[128] Powell would not be the only officer in charge of a riverine contingent. In the last four years of the war, Lieutenant John McLaughlin commanded more than 150 small boats, canoes, and barges along with more than 600 men.[129] This was the American Navy's first foray in extensive, long-term riverine operations—especially with steamships. That 2 junior officers were the key personnel for this mission represents a turning point. A new generation of naval officer charted a new and necessary course for the Navy. These technologies and operations separated them from the senior, conservative naval officers under whom they had previously served. The same cultural separation would occur with the Marine Corps.

Four days before the Dade Massacre, Marine Corps commandant lieutenant colonel Archibald Henderson had written to Dickerson opposing the administration's proposal to integrate the Marine Corps and Army. Henderson won that fight, but the Second Seminole War presented a challenge. Jackson needed more ground forces in Florida; he found them with the Marines. Jackson ordered Henderson to gather as many Marines as possible and take them to Florida where they would report to Army general Thomas Jesup. It is unknown if Jackson was simply mobilizing more forces or if he was using this opportunity to try to integrate the two services as he had first hoped. Regardless, Henderson raised two battalions and conducted operations in Florida. Confident that it would be a short war, he wrote to Secretary of the Navy Dickerson, "The new movement of the Indians may yet detain me two or three weeks in Florida."[130] The war would last another five years.

Two months after Jackson's farewell address in 1837, Henderson wrote to Dickerson about his assessment of joint operations but more so about the integration of Navy and Marine commanders:

> The naval officer is no soldier, and should not therefore command a branch of the Service of whose organization, drill and appropriate discipline they know nothing. It is in vain that they agree that because they command at sea, they should also command on shore. . . . It is essential that the stations for its drill and discipline must be under military control and commanded by military officers. It is my firm belief formed from more than thirty years experiences, that the utility of the Corps as well as its military efficiency will be fatally affected should not this be the case.[131]

Two weeks later, Henderson and other Marine Corps officers returned from their deployment by way of the steamer *Merchant*. Like their riverine counterparts, Henderson's new generation of Marines operated in an environment and for a period that had been previously foreign to them. Marines were used on ships and for occasional shore parties under the leadership of naval officers.

During the First Seminole War, Jackson completed military operations within a year (1817–18). It is possible that he thought that with the power of the presidency the result would be the same or that he could defeat the Seminoles in less time. As the war progressed late into 1836 with no end in sight, a new presidential election took place. On December 5, 1836, Jackson delivered his annual message to Congress. Much of it addressed the war in Florida. He noted the hostile aggressions of the Seminoles and Creeks required an overwhelming response by most of the Army, Marine Corps, and militias. Early successes in defending coastal settlements had not sufficiently weakened the tribes. He argued that the executive branch required additional appropriations to continue the war, while touting the patriotism of the soldier. He recommended funding for additional benefits for the soldiers as well as for private citizens whose properties the military had required.[132]

Congress had little choice but to continue funding the war and the troops. Nevertheless, it did not fail to criticize the president or the generals. Members called for investigations into the war's delays and failures as well as its growing cost. Two generals were court-martialed as General Clinch declared that the blame lay instead at the feet of the War Department.[133] Massachusetts congressman Caleb Cushing declared the war "a blot, a shame on the national reputation. . . . The blood of our people has been wasted and squandered in those arid sands."[134]

What do naval operations during Jackson's administration demonstrate? Jackson gave clear direction to the Navy in what he sought to do globally. First, he would expand American trade opportunities throughout the world. He did so in Europe, South America, the Middle East, and the Far East. Second, he assured merchants that he would protect them. By modifying the force structure of squadrons, he was able to respond to changing threats. Third, he promised a decisive response to any aggressors if American commerce was threatened. But it is important to understand that Jackson's assurance of punitive action was not his only choice. In the *Awashonks* and *William Penn* incidents, for example, American sailors may have been attacked and killed, but the incidents occurred in small, insignificant islands that bore no relation to American commerce. Other incidents, such as Falklands and Sumatra, however, had strategic or economic implications.

Two historical theories have emerged from Jackson's approach to the Falklands and Sumatra crises. The first suggests that Jackson was impetuous and punitive. One historian argues: "Old Hickory's impetuous and pugnacious diplomatic style made predictable his use of the Navy to respond to outrages on American commerce. . . . In response to separate incidents at

Quallah Battoo in Sumatra and the Falkland Islands off the coast of Argentina, Jackson dispatched warships to retaliate promptly rather than seek diplomatic solutions to attacks on American trade."[135]

The evidence, however, asserts the contrary about Jackson's ability to handle a crisis. Another historian states that "virtually all of the authors who touch upon [the Falkland Islands] crisis have ignored [Jackson's] involvement."[136] In the case of the Falkland Islands, Jackson was not reacting to a crisis. He had anticipated it. Based on newspaper reports, Jackson had knowledge of the Argentinian claims and reports of possible attacks. He sent a new charge d'affaires to Buenos Ayres specifically to deal with this situation, and he supported diplomacy with appropriate naval forces in the region. He also ceded control of the Falklands to the British rather than enforce the Monroe Doctrine. In his fourth annual message to Congress, Jackson refrained from providing too much information about negotiations with Buenos Ayres, which would be inexpedient, thereby again avoiding the nationalist fervor that might have erupted and been exploited by another president.

In the case of Sumatra, Jackson's orders to Downes were clear. Downes simply did not follow them. Jackson wanted compensation and wanted the guilty Malays brought to justice, and he recommended the use of force only as a final option. Downes attacked as his first option. Despite the turmoil of his administration's resignations, Jackson demonstrated a thoughtful, reasoned, diplomatic, and forceful approach to international crises. His vision went beyond his stated goals of enhanced trade, protection of commerce, and responsive action. A decade before the term "Manifest Destiny" became the rallying cry for Americans to colonize the West and push to the Pacific Coast, Jackson was pursuing his own maritime destiny for the nation that connected the United States with the rest of the world.

Asia appealed to Jackson as part of his effort to expand American trade. The presence of Navy ships in the eastern Pacific would be insufficient to meet that goal. Consequently, Jackson established the East Indies Squadron under Commodore Edmund P. Kennedy. Comprised of the *Peacock* and the *Enterprise*, the squadron reached the Arabian peninsula in September 1835. Aboard *Peacock* was Jackson's first envoy to the east, Edmund Roberts, who reached trade agreements with Muscat, Oman, and Thailand. Though tasked by Jackson to open trade with Japan, Roberts died of dysentery when the ships pulled into Macao. No previous president had envisioned the extent of naval service as Jackson did during his administration.

This was evident with the East Indies Squadron.[137] American ships must traverse that ocean to Jackson's new Asian partners. Those ships had to understand their environment—the islands at which they could resupply or make repairs, the dangers of surrounding reefs and shoals, and the increased presence of American merchants, Navy agents, and consuls. This was why Jackson ultimately endorsed the exploring expedition. Like the merchants of the Northeast, he understood that America's economic future lay not only with its traditional European trading partners but with new partners in the East. It is not known if Jackson read Alexis de Tocqueville's work, but Jackson would have been in full agreement that America was born to rule the seas. That vision of a naval future rested not only in Jackson's hands but also in the hands of a group of officers and literary figures who would articulate that vision and help Jackson carry it out.

Amid these operations, the Navy was gaining unprecedented experience in an abbreviated timeframe. The squadrons had seasoned commodores who had fought in three wars and now were vigilant in keeping the peace with other nations. Junior officers had a new technology, largely untested in combat, thrust upon them as they planned and coordinated riverine squadrons. This would later prove invaluable during the Mexican-American and Civil Wars. But they also did this in a decade that changed the professional culture of the Navy.

6

An Actualized Naval Culture

The Jacksonian era seemed to sweep up all aspects of American life. A system of naval personnel that had remained stagnant since the founding of the republic was no different. Senior officers had remained in service for decades, often carrying conservative thinking through to the 1830s. The Navy was developing its own unique culture and a formalized structure to better develop expertise as a modern Navy.

Underlying this professionalization was accountability and that was evidenced by courts-martial. The sixteen thousand pages of records from more than three hundred courts-martial during Jackson's presidency provide unique insight into the challenges of discipline, infighting, and morale. Officers were held responsible for their actions. While this was true for any period, Jackson's personal intercession with so many proceedings reinforced a broader message to the Navy—the president expected discipline among the ranks to improve.

The components of professionalization occurred concurrently. Junior officers resented senior officers as a group since their resistance to retire left few advancement opportunities and they were able to make some inroads in changing the rank system. Naval schools at a few stations, such as New York, Philadelphia, and Norfolk, were deemed insufficient with the introduction of new technologies and formalized colleges and universities established across the country. Therefore, the officers proposed a new, centralized shore-based academy.

Advancements in medicine led to the establishment of a medical corps. The Second Great Awakening resulted in a formalized Chaplain Corps. Alcohol, such a vital component of shipboard life because of water purification challenges, was the cause of discipline problems, but the administration tried to limit its use. Finally, Jackson initially sought to incorporate the Marine Corps with the Army, but senior naval officers convinced the president of its utility and that it deserved a more enhanced role within the Department of the Navy. Taken in total, these components represented a fundamental shift toward naval professionalization.

Courts-Martial

Few aspects of the Navy were more important than discipline, especially at sea, where precision, cohesion, and effort were necessary to survive the natural elements that might sink a ship and enemy engagements, which could do the same. Without strict naval discipline, anarchy and lethargy ship-wide might not reign supreme, but it only required one weak link in the chain to diminish the fighting or sailing capability of the ship. Courts-martial records provide a wealth of information about the officer corps, operations, and the technical and social aspects of the Navy. The content in those records also tells the story of the president's own role in naval justice.

The survey of US naval courts-martial and courts of inquiry from 1829 to 1838 illuminates several issues during the Jacksonian era. On average, thirty courts-martial or courts of inquiry occurred annually. While the number generally ranged from a low of ten proceedings in 1831 to thirty-eight in 1830, a spike occurred in 1835 with eighty-two courts-martial. Twenty-five of these proceedings were for sailors and officers; the remainder were for Marines. Although Marine officers and noncommissioned officers had been court-martialed prior to 1835, an act of Congress only recently included Marines within Department of the Navy for courts-martial. Unlike the Marines of the late twentieth and early twenty-first century, the Marine Corps of the 1830s was often the military branch of last resort. Commodore Charles Stewart, who had served in the Quasi-War, the Barbary War, and the War of 1812, told his long-time friend and diplomat Richard Rush in 1817 that the Marines were the dregs of society but necessary for ships.[1] Most of the Marines charged for violations were court-martialed for desertion, drunkenness, and insubordination. The significant drop in prosecutions after 1835 suggests that an intensive effort to cull the worst violators had been largely successful.

Marines were also the greatest percentage of offenders in shore commands. In Charlestown, Massachusetts, and Norfolk (including Portsmouth and Gosport), Marines accounted for 46 percent and 42 percent respectively of those officers and sailors charged with court-martial offenses. In Brooklyn, they accounted for over 66 percent. And in Philadelphia, Marines accounted for an astounding 92 percent of all court-martial proceedings. Most Marine Corps courts-martial occurred in 1835, a reflection of the Marine Corps Reorganization Act of 1834, which subjected Marines to Navy regulations. This is supported by the number of naval officers court-martialed compared to their years in service. Most courts-martial for officers between 1829 and 1838 occurred in the first six years of service (fig. 6.1). The number of charges by rank alone (exclusive of specific years in service) bears this out. Of the 106 courts-martial of naval officers between 1829 and 1838, 51 were of midshipmen. Of the 80 courts-martial of sailors, ordinary seamen were involved in 54.

When officers and sailors were at sea on long deployments, the chances officers and sailors would commit an offense was minimal compared to when based on shore in the United States. In fact, nearly half of all courts-martial proceedings occurred in one of the four major established shore commands

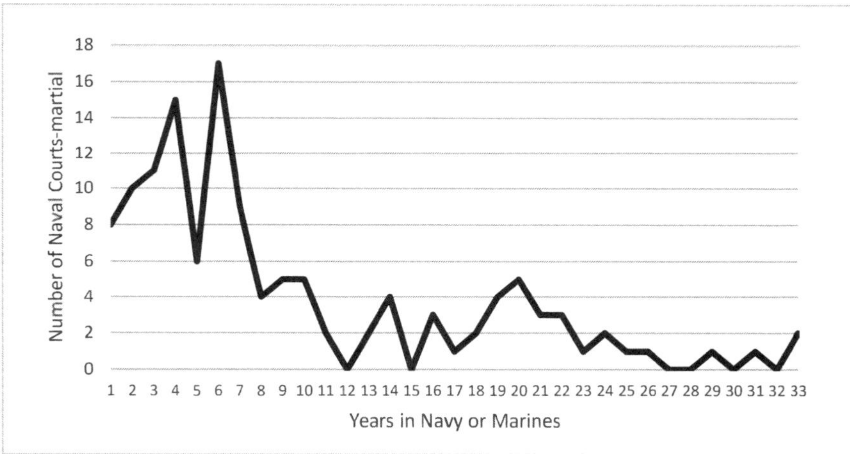

6.1. Navy courts-martial, 1828–38, by years of service. Compiled by author from Records of General Courts-Martial and Courts of Inquiry of the Navy Department, 1799–1867, Navy Courts-Martial, 1825–1840, Record Group 125, M273, National Archives.

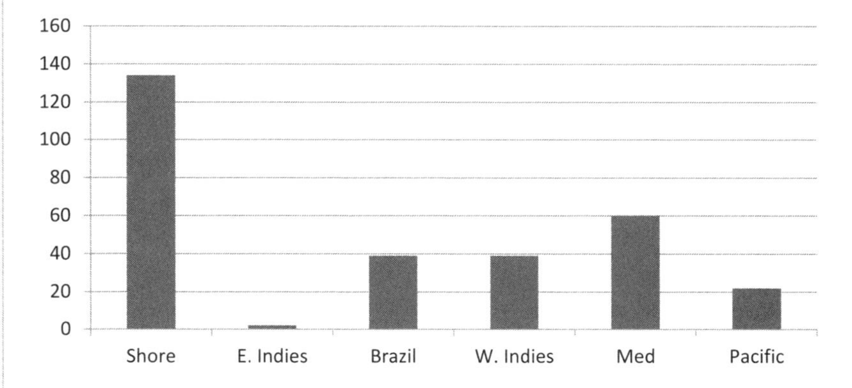

6.2. Navy courts-martial, 1828–38, by station. Compiled by author from Records of General Courts-Martial and Courts of Inquiry of the Navy Department, 1799–1867, Navy Courts-Martial, 1825–40, Record Group 125, M273, National Archives.

in Norfolk, Philadelphia, Brooklyn, and Charlestown. Of the remaining incidents, the Mediterranean was the hotbed of activity, perhaps for the simple reason that the greatest number of sailors was based with that squadron (fig. 6.2).

Since the promulgation of the Naval Regulations of 1802, little had changed in terms of expectations or consequences for violating required behavior. Despite clear regulations, sailors and officers violated the rules. Between 1828 and 1838, 310 Navy courts-martial of sailors and officers produced over sixteen thousand pages of documented proceedings. Not every sailor charged was given a court-martial—only infractions meriting twelve lashes or more resulted in a court-martial. Officers, unlike sailors, were subject only to courts-martial, and any finding of guilt could not result in corporal punishment.

A curious inconsistency developed out of this atmosphere characterized by a code of iron discipline that imposed the strictest possible control over the enlisted men. Because officers often regarded them as the lowest class of humanity, their punishments were mainly flogging, confinement in irons, and the threat of capital punishment, claiming that was all they could be expected to comprehend.[2]

Although corporal punishment through the cat-o'-nine-tails or irons were a common punishment for sailors, more draconian measures such as hanging a man by the yardarm largely occurred only in myth rather than the Navy. When Landsman Thomas Allen was charged and found guilty of murdering a seaman on the island of Minorca, the court "sought with anxiety to establish a plea of insanity in behalf of the Prisoner of some mitigating circumstance in his favor."[3] The court hoped the crime was committed because the accused was intoxicated. Finding no mitigating circumstance, it reluctantly sentenced him to hang. In another rare case of the death penalty, a sailor was charged with cutting the throat and causing the death of another sailor. He was found guilty and sentenced to be "hung till he be dead at the yard from the *Boxer* at Sea."[4]

The first man executed in the Navy was a seaman who had been found guilty of cowardice on the USS *Constellation* during its 1799 engagement with the French *L'Insurgente*. According to one historian, between 1799 and 1862, only twenty-six sailors were sentenced to death, with eleven confirmed and six probable death sentences.[5] Only one officer was put to death—Midshipman Philip Spencer, the son of the sitting secretary of war, was summarily executed aboard USS *Somers* in 1842 for fomenting a mutiny. In 1831, Secretary Woodbury issued a circular recommending flogging as the last resort. While some Navy captains, like Charles Stewart who commanded ships from 1800 to 1841, sparingly flogged sailors, flogging continued until 1855.[6] In 1846–47 alone, for example, there were 5,936 cases of flogging.[7]

Nevertheless, excessive use of the lash was rare. Between 1800 and 1851, only forty-nine cases were given sentences of a hundred lashes or more. In 1837, one seaman, John Herring, was found not guilty of murder but guilty of manslaughter. The seven-member court's sentence was four hundred lashes, the most meted out by any court.[8] In this case, Commodore Jesse Duncan Elliott remitted the sentence to three hundred lashes, with one hundred lashes to be administered on each ship of the squadron—the USS *Constitution*, the USS *Shark*, and the USS *United States*.

Courts-martial were conducted by the officers themselves as prosecutors, jurors, and judge. Three or more officers sat in judgment. No officer could be junior to the officer or sailor prosecuted lest they be intimidated by the defendant's rank, if senior enough. In addition, when the board publicly stated their vote on a case, the most junior officer on the board spoke first in order that the board would not change its vote upon hearing the decision of a more senior member of the board. One author argues that the justice dispensed by this system was often irregular and clearly incompatible with the system of civilian law. "It had to deal with problems arising from cronyism, factionalism, grudges, and feuds that festered deeply because the participants could not escape one another in the enforced intimacy imposed by a small fleet with a limited number of berths."[9] The author further contends that "since nearly every senior officer cultivated political contacts or at least could rely on the congressional delegation of his home state to back him, political pressure was quickly mustered to set aside either convictions or sentences."[10] While some of the proceedings were conducted at naval stations, the author makes this assertion with little documentation to support it. Courts-martial often took place on ships that were weeks, if not months, away from political support or interference.

The last aspect of violations is categorized by ship. The 212 courts-martial occurring on ships resulted in an average 5.7 proceedings per ship; this meant that for the period of 1829 to 1838, each ship averaged less than 1 court-martial per year. Ships of the line, with over 700 crewmen, should have had a higher average of courts-martial simply because of the higher number of crew and opportunities for incidents. Instead, the five ships of the line (USS *Independence* had not yet been converted to a razee frigate[11]) averaged 4.8 courts-martial per ship, or less than 1 every two years during this period. The most likely reason for this particular anomaly was better discipline, since ships of the line were commanded by or had on board the squadron commander, a captain who had probably first been commissioned during the Quasi-War or the Barbary War. In either case, these captains and even lieutenants might have had several decades of experience each and simply possessed better leadership and management skills.

The Jackson administration's expectations for accountability and discipline were constant, but the number of courts-martial was not. Except for 1835 and the inclusion of Marines in Navy regulations causing a spike of eighty-five courts-martial, the annual number of courts-martial ranged from twelve to thirty-six.

The seminal naval court case of Jackson's administration was the inquiry into Lieutenant Robert B. Randolph for two reasons. First, Jackson's reviews of court cases up to that point had been measured, informed, and lenient. The Randolph case completely contradicted the aspect of Jackson's judicial background. Second, the case manifested Jackson's growing overly emotional and irrational reaction to the Petticoat Affair.

Margaret Eaton, as readers will recall from a previous chapter, was married to John Timberlake who gained a position on the USS *Constitution* as purser, responsible for accounts. Timberlake died in April 1828 and was buried in Minorca. The ship's commanding officer, Commodore Daniel Patterson, ordered Lieutenant Robert Beverly Randolph, an experienced officer who was commissioned in 1810 but had no experience with accounts, to assume Timberlake's responsibilities. When the ship returned, the fourth auditor of the Treasury, Amos Kendall, began questioning the accounts.

On June 4, 1832, a court of inquiry was convened at the Charlestown Navy Yard with Commodore Charles Morris presiding. The inquiry's findings were reported to the secretary of the Navy on January 19, 1833. Randolph was charged with financial irregularities. Decades later, Randolph's daughter claimed that her father requested a court of inquiry to clear his name. For unknown reasons, Secretary of the Navy John Branch initially denied Randolph's request. The request was not granted until Levi Woodbury succeeded Branch. The court of inquiry, however, assessed that Kendall requested the proceedings.

In trying to defend himself, Randolph had sent Timberlake's steward to the office of Fourth Auditor Amos Kendall to examine the accounts. Timberlake had left approximately $11,000 in cash (including gold and silver) at his death aboard the *Constitution*, and this had been transferred to Randolph as acting purser.

Complicating the issue was the question of vouchers claimed to be paid before and after Timberlake's death. These included a considerable number of vouchers when the ship pulled in to resupply at Gibraltar before returning to the United States. Randolph then settled the *Constitution*'s accounts in

Boston upon its return in 1828. He traveled to Washington to meet with and deliver the account books to then-Fourth Auditor Tobias Watkins, who himself had been tried in 1829 for fiscal mismanagement. Decades later Randolph's daughter recounted, in her father's defense, that he had received official notice from both Watkins and the second controller that the account was closed and certified.[12]

In his position as fourth auditor, Kendall had copies of everything filed but denied both Timberlake's steward and Randolph access when Randolph first received notification of discrepancies in 1830, which resulted in more than a few words by Randolph, who was adamant about ensuring full accountability. One problem was that Timberlake's account book disappeared. As one historian notes: "The tale becomes more complicated when the revelation appears that Senator John Eaton (soon to become Secretary of War under Jackson and husband to Timberlake's widow [Margaret]) was the guarantee of Timberlake's bond. If the dead purser had defaulted, the financial responsibility was Eaton's. In a senate hearing held in February 1830 with treasury information provided by 'Kitchen Cabinet' members Amos Kendall as Fourth Auditor and Isaac Hill as Second Comptroller the Senate exonerated Timberlake (and therefore, Eaton) from culpability. Denied access, Randolph said he would apply to friends in the Senate (specifically John Tyler and Littleton Tazewell) to obtain a fair accounting."[13]

After an extensive investigation, the court criticized Randolph for making payments but not taking receipts. It found that he had not taken proper inventories but acknowledged that he had received only verbal orders—not the normative written orders—which may have been misunderstood. There was no evidence that Randolph had deceived his commanding officer about the money or supplies. The court found that it could not account for $4,303. In the end, the court agreed that Randolph's irregularities in his manner of performing the duties of acting purser were not an intentional effort to defraud the government.[14] "It does not appear," they concluded, "that Lt[.] Randolph said anything with any intention to mislead or deceive the commanding officer."[15]

That should have been the end of the case, particularly since it was a court of inquiry designed to investigate the facts of an issue, as opposed to a court-martial, which carried with it the possibility of judicial punishment. After a three-year fight to clear his name, Randolph was clearly relieved to be exonerated, but that relief turned to disbelief when Jackson himself provided an additional ruling on the case.

The fact that only the commodore of the squadron, the secretary of the

Navy, or the president could change the outcome of the proceedings of a court-martial was one of the hearts of the issue during the Somers Mutiny a decade later when the commanding officer, Alexander Slidell Mackenzie, hung three crewmen (including the son of the sitting secretary of war) absent a trial or approval had the trial occurred. In most rulings, Jackson reduced the sentencing. His activist approach likely originated from his sense of frontier justice and his own experiences as a lawyer and jurist. Most presidential rulings found in the court-martial records are terse—usually a few sentences to state whether Jackson approved, disapproved, or modified the punishment. This was not the case with Randolph.

Stunningly in April 1833, Jackson offered a detailed four-page reversal of the court of inquiry. Jackson pointed out that although the money was counted, no inventories of the stores were conducted. From the records and files of the fourth auditor's office, it also appeared that neither Randolph nor anyone else furnished that office with an account of the money or the stores. For this and other issues he discussed, Jackson wrote, "The President can not approve of so much of the finding of the Court as declared."[16] Nor could he approve of the court's opinion that Randolph's irregularities were not intended to defraud the government.

Jackson took an unfairly heavy-handed approach by stating that Randolph was "unworthy of the naval service of this Republic, and an unfit associate for those sons of chivalry, integrity, and honour, who adorn our Navy."[17] Jackson ordered the secretary of the Navy to dismiss Randolph, again, an action Jackson was not authorized to do absent a court-martial. Further, he wrote: "The President trusts that the most efficient means will be resorted to by the Navy department to prevent in future the total neglect and disregard to deceased officers and their families which form striking characteristics in this case."[18]

A few weeks later, Jackson, Donelson, and others boarded the steamship *Sydney* in Alexandria bound for Fredericksburg, where Jackson would participate in a ceremony honoring Washington's mother. *Sydney* returned on May 6 to Alexandria.[19] A company of Marines was also on board. Jackson sat alone in the salon, smoking his pipe and reading a newspaper when Randolph entered. Jackson rose to extend his hand toward the stranger when the now-former naval officer announced himself as the naval officer whom Jackson had wronged and struck him. Some accounts state that Randolph tweaked his nose, or rather, "the man grabbed Jackson by the nose and proceeded to turn it until blood flowed in a steady stream."[20] Randolph's action thus got recorded as the first attack on a sitting US president.

Several Jackson supporters tried to stop Randolph. He escaped to Richmond and the safety of his family. That evening, the prominent author Washington Irving spent the evening with Jackson and later recounted to Irving's brother: "The old gentleman was still highly exasperated at the recent outrage offered him by Lieutenant Randolph. . . . It's a brutal transaction, which I cannot think of without indignation, mingled with a feeling of almost despair that our national character should receive such crippling wounds from the hands of our own citizens."[21]

For months, Randolph found himself the subject of newspaper articles, first for the court of inquiry and then for the attack on the president. Some supported him in both while others admonished him. The support and attacks tended to arise from the highly partisan nature of the newspaper industry, many of which were simply political organs of the Jacksonians or Whigs. One example was the *Globe*, which Jackson established as the newspaper of record for his administration.[22] The editor was Francis P. Blair, a Jackson stalwart who had helped him win Kentucky in the 1828 election and who became part of the president's "Kitchen Cabinet." While in Kentucky, Blair had been a writer for the *Argus* of which Amos Kendall was the part owner and editor in chief.

The *Globe* was relentless in its defense of Jackson by attacking Randolph and the Navy. Two days after Randolph's assault, the paper proposed that the court of inquiry itself was a sham because the members were "persons very favorably disposed toward the accused."[23] With the liberty of full, front-page spreads, the *Globe* kept its attention on Randolph for having verbally challenged Kendall and physically attacking Jackson: "It is fair, too, to presume that gentlemen of the sword being wholly inaccessible to fear and incapable of being swerved from their duty by the threats of desperado, cannot conceive that threats are ever intended to embarrass or intimidate public functionaries."[24]

A warrant was issued for Randolph's arrest. When he was found later that year, John Marshall, the chief justice of the Supreme Court, heard the case in the circuit court. The court found the warrant was unconstitutional under an 1820 law, and Randolph found himself a free man—free but not forgotten. "The act maimed his prospects, and he lived and died in poverty."[25] He had been commended by Stephen Decatur for actions in the War of 1812, but following the court of inquiry and attack on the president, Randolph was dismissed from the Navy he had served, attacked in the press, and left to a life without a mission.

Within the context of naval justice during this decade, the Randolph case is important because of how Jackson followed this and overreacted. He had no authority to dismiss an officer based only on a court of inquiry. Jackson's motives in Randolph's dismissal are likely due to his obsession and the personnel involved—John Eaton, Margaret Eaton, and Amos Kendall—rather than the procedures. Like most issues in Jackson's first term, the Eaton Affair clouded the president's judgment, and a naval officer simply became the target of one of the subjects of his anger. The Randolph case is significant from another perspective.

Jackson involved himself with a quarter of all Navy courts-martial. His involvement was largely consistent and regular—except for a two-year period following Randolph's attack in which he commented on only two cases. The cause of this discrepancy cannot be determined through Jackson's correspondence or through other sources. It is possible that Jackson recognized he had overstepped his constitutional authority or even felt remorse over the decision, a trait not normally attributed to him by biographers. Perhaps he reflected that his decision had indeed been heavy-handed and unfair. If there is any evidence to support this, then it would be a later letter from Jackson to his successor, Martin Van Buren, stating that if Randolph were ever found guilty to give him a presidential pardon.[26]

Jackson appears to have viewed his role as a check on a service within his own branch of government as indicated through naval courts-martial. Several officers presided over courts-martial. Of the 271 courts-martial during his two terms in office, Jackson approved, mitigated, or overturned 65 of them; his secretaries of the Navy overturned an additional 67. Compared to other presidents from John Quincy Adams through James Buchanan, Jackson directly involved himself in more than twice the cases than any other chief executive.

Education

Education was key to developing a professional officer corps as well as advancing the Navy technologically and strategically. Although the Navy did not establish a formal naval academy until 1845, the prior decade saw many incremental changes as well as wholesale proposals. Reforms in the selection of midshipmen was crucial to this gradual professionalization. According to one historian, "many midshipmen were perceived as incompetent, reducing efficiency and increasing expense. Incompetence was attributed to appointments based on patronage requirements rather than ability, and inadequate opportunities to learn their profession."[27] This is true—patronage was the primary source for commissions. Louis Goldsborough was only six years old when he received his warrant (his father had been the Navy's longtime chief clerk). Others, like John C. Calhoun Jr., the vice president's son, were from prominent political families. Charles Steedman received his warrant in 1828 because his father knew the president pro tempore of the Senate.

In the last few months of his term in office, Adams's secretary of the Navy, Samuel Southard, asked Congress for a reorganization of the Navy and Marine Corps, a criminal code, an increase of ranks, a naval school, education of American seamen, and provisioning of naval hospitals. It was not the first time such a proposal had occurred, especially regarding education. In 1829 Branch reported that the subject of schools had often been recommended and failed to pass so often that elected officials were reluctant to continually vote on it. Again, Jackson did not have to present this to Congress. That he did so suggests an intention to carry forward previous administrations' policies. Perhaps, like Branch, he recognized the need for an educated naval officer corps to conduct diplomatic negotiations globally. Part of that reflected Branch's belief that the Navy ignored foreign languages because they were viewed as a great inconvenience.

The Navy had two schools—in the Navy yards in New York and Norfolk—for mathematics since that was required for navigation. When the future rear admiral Charles Steedman became a midshipman, he reported in 1829 to the Naval School at the New York Navy Yard commanded by Commodore Isaac Chauncey. The head teacher was Edward C. Ward, professor of mathematics in a school of thirty to forty midshipmen.[28]

Perhaps recognizing the shortcomings of the naval schools, William A. Duer, president of Columbia College, offered classes at the school to local young officers.[29] Chauncey considered it. "This proposal," Chauncey wrote to Branch, "is a liberal one, not more expensive than the Navy yard schools. I certainly should prefer a Naval School, if Congress would authorize one."[30] If the Navy accepted, Chauncey believed that a naval officer could be attached to the college for "superintending the young officers, and enforcing discipline." However, there is no record at Columbia University of any midshipmen attending classes in the 1830s.[31]

In Branch's report to Congress in 1830, he noted that naval officers ought to study more than mathematics, including foreign languages and international laws. "The sons of the wealthy," he wrote, "may obtain these advantages from the bounty of their parents; but without the aid of public instruction, how are the sons of the less affluent to become qualified to command in the naval service?"[32] For more than two decades, Army officers had been educated at West Point. Aside from the two math schools in Navy yards, the only provision for the education of midshipmen was a schoolmaster and, even then, only on larger ships.

The naval schools in New York and Norfolk of the 1830s were a crucial step toward the establishment of a formal, common school. The fundamentals leading to its establishment existed. Naval officers who traveled around the world understood the need for a standardized, shore-based curriculum by professional instructors, particularly in the sciences and engineering.

It is important to note the significant difference of the two streams of thought that finally led to the Naval Academy. Naval officers campaigned for a school out of a need for technological advancement to understand steamships and how they would be used operationally. By contrast, Congress was responding to a single incident with the politically and emotionally driven call for a standard school. It was the officer corps that understood the cold, rational necessity of advancing the profession. While an academy was still a few years away, the professionalization of the staff corps was well underway in the 1830s.

In February 1837, just two weeks before Jackson's term expired, Samuel Southard introduced out of committee Senate Bill S. 64, "A Bill to Establish a Naval Academy," based on "the earnest recommendation of the Executive."[33] In 1836, naval officers—largely lieutenants, midshipmen, chaplains, and surgeons—petitioned Congress for a naval academy.[34] They pointed out that the school master system was ineffective because school masters could rarely impart elementary or scientific knowledge or advance the education of naval officers. They argued for a naval school that could "help them prosecute the profession to which [they were] devoted and perpetuate the commercial prosperity of the country entrusted to their safekeeping on the high seas."[35] They asked that their petition be presented to the secretary of the Navy.

This petition was significant. Each petitioner signed his name to the document, unlike the pseudonymous articles in the decade's magazines. They chanced retribution from senior officers and the secretary of the Navy. None, however, appear in court-martial records. More notably, of the thirty-four lieutenants and midshipmen who signed the petition, all but five were promoted (one having died in the interim and the other four resigning later in the decade).[36] There was no retribution for addressing the issue to Congress and challenging the schoolmaster system. Nevertheless, Southard's bill narrowly lost the vote on March 3, one day before Jackson left office.

It would, however, take an incident to convince Congress and the public of a need for a school that taught ethics and morals. In 1842, the commanding officer of the brig *Somers*, Alexander Slidell Mackenzie, uncovered what he believed to be an imminent mutiny. Without a trial or sanctioned by the Department of the Navy as the ship was weeks from shore, Mackenzie ordered

Acting Midshipman Philip Spencer (the son of the sitting secretary of war) and two sailors to be hung. When the ship returned, Mackenzie was tried in a court-martial covered in the papers by such luminaries as James Fenimore Cooper and Richard Henry Dana, author of *Two Years Before the Mast*. The trial called attention to how midshipmen were educated and resulted in the establishment of a naval school in 1845 at the old Fort Severn in Annapolis.

Medicine

Medicine and the establishment of a professional Medical Corps reflected scientific advance and the willingness of at least some officers to improve and modernize the Navy.[37] While imperfect, the Medical Corps represented a new professionalism, uniformity in education and training, as well as recognition of the importance of the overall health and well-being of officers and sailors. Thus, the Navy recognized the need for qualified and quality surgeons and proper medical care. Two primary determinants influenced the role of medicine and the treatment of disease as a component of naval culture during the Jacksonian era. The first were geographic and operational changes. The second had to do with the gradual professionalization of the medical profession. It was a pivotal decade for the naval medical profession, as the Navy founded naval hospitals throughout its major bases and commissioned more qualified surgeons, while Congress recognized this need by increasing the medical portion of the Navy's budget by 50 percent.[38]

Surgeons in the antebellum Navy assumed the roles of physicians, psychologists, and coroners during ship deployments, and quite often, they failed to excel in any of those duties. Far more often, sea-based surgeons were a pale shadow of the capabilities of their counterparts on land. Medicine had yet to benefit as much as other sciences from the Age of Enlightenment. Bloodletting, for example, was still a widespread practice in the mistaken belief that it represented a reliable relief of symptoms and disease.[39]

William Barton, like many Navy surgeons, was from Philadelphia, a center of American education and affluence. He advocated the use of lemons and limes to fight scurvy, composed regulations for governing naval hospitals, and promoted the employment of female nurses.[40] His arguments occurred in the late 1820s, decades before Clara Barton played her role in the Civil War.[41] Many diseases resulted from the proximity and tight living quarters on ships at sea. Unlike Army soldiers, who when not in barracks were often in the field, Navy sailors and officers had nowhere else to go. Ventilation was not a problem with the Army, whereas on ships it was extremely poor and a contributory cause of disease. Captain James Barron attempted to rectify this in the early 1830s with his patent for bellows, which could ventilate the confined lower decks where the crew ate, slept, and fought.[42]

Contemporary medical discoveries reduced deaths in the fleet. Two such examples were the vaccination for smallpox and, increasingly important to the Navy, the use of quinine to treat malaria (1820).[43] By the 1830s, the Navy had established squadrons off West Africa, the West Indies, and the Gulf of Mexico, all of which exposed sailors to tropical diseases. This represented a dramatic operational shift that had direct implications on the ability of personnel to perform their duties and the costs associated with widespread disease. Some physicians and commanders recognized the need to adapt to these changing conditions. Matthew Perry, for example, commanded the *Shark* off West Africa and instituted the "Rodgers system" to keep sailors healthy, keep them hard at work, and minimize shore leave.[44] This may have been the result of the *Peacock*'s experience, which in June had found itself plagued by yellow fever that claimed the lives of four officers and several crewmen. That ship's experience clearly had an impact on the morale of at least some other officers. Lieutenant Garrett J. Pendergrast had concerns about his fellow officers deployed to hazardous regions. "I have read with much regret the accounts given of the fatal consequences produced by the yellow fever in our W. India Squadron." Pendergrast thought that fellow officer Samuel Du Pont had been "numbered with these gallant fellows who have sacrificed themselves for their ungrateful country's good."[45]

Prior to Jackson's administration, naval hospitals dispensed minimal medical care. Patients found themselves receiving tepid care and being ignored, at best, or maltreated, at worst. Dr. John Wiley of the naval hospital at Brooklyn was the cause of numerous egregious abuses and rampant misadministration.[46] Seaman Charles Brooks died after another patient administered three tablespoons and one teaspoon full of laudanum to him. At the time of the seaman's death, no one was awake in the ward or knew of Brooks's death until another patient awoke in the morning. Seaman Samuel Brown was put in double irons on November 13, 1824, for nearly two weeks and then received one dozen lashes on his bare back by Wiley's order. Wiley then had Brown put in double irons again, this time with his hands behind his back for four more days.[47] After his lashes, Brown failed to receive any medical treatment. Seaman Ezekiel Whitney begged for someone to wipe his eyes "that he might once more be enabled to behold light" before his death from smallpox. With no one appointed for that purpose, a Marine corporal tended to him.

Another sailor, in the service for over twenty years, found himself tied up and flogged while he begged Wiley for mercy. The same was true of another sailor punished with twelve lashes. These were just a few of the patients flogged or threatened with it. Marine corporal John W. Jones testified that, when he was a patient from July 16 to September 2, 1824, "no one was appointed to attend the sick as a nurse in any wards. . . . One patient was obliged to wait on the others or have no attendance." He also stated that "the attending physician never visited the wards oftener than once."[48] One of the surgeons, Dr Mordecai Morgan who was affiliated with the hospital for only two months, testified that he supported the concept of patients caring for themselves as well as lashing.

Moreover, despite the inadequate care, there was a general lack of discipline among Navy surgeons. Dr. John Kearney, surgeon of the fleet in the Mediterranean, found himself charged in May 1829 with disobedience of the lawful order of his commanding officer, specifically staying overnight in the Port of Mahon on several occasions. Commodore William Crane had directed that no officer should remain away from the ship overnight.[49]

Kearney then wrote to Crane asking if the order applied to him, since he had been "visiting the sick ward frequently enough . . . during the days" and had assistant surgeons observing the sick when he was not present. Crane was clear in his response: "Your situation is not one of convenience but responsibility, and whether it is your duty personally to attend to the prescriptions or not, I believe it to be necessary and proper that you should see them administered." Therefore, Kearney was not to sleep away from the ship especially since one crewmember had died without the presence of any medical officer. The following day Kearney responded, oblivious to the implication of the crewman's death: "But this will confine me to the sick ward most of the day and night."[50]

Kearney further expected Crane to offer him some of the same liberties offered to other wardroom officers ashore. Crane acquiesced only if the sick were free from danger. Kearney saw his opening in defining "danger." On May 15, Kearney noted that all sick personnel were free from danger and asked that he leave the ship. Crane rejected the request. Despite clear direction from the commodore, Kearney went ashore. After several days, Crane ordered him to return. Kearney claimed that he was too ill to return, confined to his bed. Unfortunately for Kearney, he had encountered Lieutenant Edmund Byrne the evening before. Byrne testified at the following court-martial that the doctor had seemed healthy enough. "He told me he was going to the Opera and that he had come out between the acts on account of the warmth."[51] A court-martial found Kearney guilty, suspended him from duty for six months, and dismissed him from the Mediterranean Squadron.

Laudanum, an opiate used to treat a variety of conditions, was used as frequently as an analgesic as for treating alcoholics. Crane charged Lieutenant James Ramage of the *Delaware* with contempt of authority and scandalous conduct, primarily being intoxicated. While changing in the wardroom one evening, Lieutenant Edmund Byrne noted Ramage's bloody shirt. The latter dismissed it as a wound he had received in the United States before they got underway. "He shewed [*sic*] me a large vial containing some liquor which he said he was about to take to alleviate the pain," said Byrne, noting that it was laudanum.[52]

The prior cases suggest a need for professionalization of the medical corps. Commodores and captains expected higher standards for surgeons by the 1830s. If not, then they at least demanded accountability from those who failed in their primary duty, which was the health and well-being of the crew. The secretary of the Navy reported to Congress in December 1830 that the Navy required special attention for the construction of appropriate facilities particularly in Pensacola.[53] The Navy had already purchased land for hospitals at the New York and Boston yards. Pensacola's mild climate would help those suffering from maladies.

In 1830, Barton published his book *Hints for Medical Officers, Cruising in the West Indies* before assuming his duties at the Norfolk naval hospital. This work was fortuitous as the Navy's presence in the West Indies and Gulf of Mexico would substantially grow throughout the decade, particularly during the Second Seminole War of 1835 to 1842. But, as diseases like typhoid, scarlet fever, dysentery, and others struck ships in the West Indies, ship captains responded. When disease struck the frigate *Brandywine*, the captain ventilated the ship in the Havana harbor. When Commodore Lewis Warrington reported the frigate *Java* had cholera aboard, the Department of the Navy directed him to stop unnecessary exposure and hard labor, keep the ship clean and dry, issue proper clothes, and distribute lime and soda.

The period after the War of 1812 meant fewer military engagements and, consequently, fewer opportunities for medical professionals to practice their craft in a battlefield environment of surgeries, amputations, and the dressing of battle wounds. There were exceptions. For example, during the murder trial of Seaman Benjamin Marsden, the Mediterranean Squadron's fleet surgeon testified with great detail and accuracy on the victim's final hours, treatment, the weapon used, and postmortem assessment.[54]

Education specifically applied to naval surgery remained limited. The

United States had few medical colleges and lagged far behind scientific education on the continent. The British had founded Greenwich Hospital in 1694, but it would take more than two hundred years before Britain established a school devoted to tropical medicine.[55] In 1823, however, Dr. Thomas Harris proposed a refresher course for surgeon's mates for promotion. Harris recognized that American naval surgeons needed to practice on cadavers as in European medical hospitals. Absent an educational facility to specifically train naval surgeons, Harris also arranged for students to attend lectures at the University of Pennsylvania. Not surprisingly, there was a paucity of qualified surgeons. As Southard wrote just a few weeks before Jackson's inauguration, "our present list of medical officers is much too small and has for several years been too small to ensure proper attention to the health and comfort of our officers and seamen."[56]

For the first time in the Navy's history, the need for a more formalized hierarchy among medical officers emerged as Senator Robert Y. Hayne (D-SC) introduced legislation on February 16, 1830, to create a surgeon general for the Navy. Upon seeing the bill's draft, Secretary Woodbury recommended that the senator water down the role to an occasional supervisor. Although the term "Surgeon in Chief" was in the original bill submitted to the committee, members changed this to "General" whose role it would be to "supervise the manner in which [medicine had] been performed, to supervise the selection, purchase and putting up of all drugs, medicines, Surgical instruments, Hospital Stores and furniture."[57] The proposed legislation ultimately failed to become law.

Four years later, Woodbury reversed himself and asked Surgeon Thomas Harris to supervise the medical affairs of the Navy. Harris was an accomplished physician who had overseen the construction of the Philadelphia naval hospital, served as president of the naval board of medical examiners, and was an acquaintance of Jackson. In 1832, Harris operated on Jackson to extract a bullet from the president's body that had resulted from a duel two decades before. Harris arrived at the White House while Jackson was in a meeting. The president excused himself and, without the benefit of anesthesia, submitted himself to the skilled surgeon who found the ragged, flattened ball lodged in an arm muscle; Jackson had suffered lingering symptoms through to his shoulder and back. Within a few hours, Jackson was at dinner with friends while his arm hung in a sling.[58] Harris would later serve as the second chief of the Bureau of Medicine and Surgery.

One shift by the 1830s, due partly to reforms in the previous decade, was the demand by naval doctors and surgeons for equality of rank and pay with line officers as well as their peers in the Army. Southard had, by 1824, established examinations for surgeon's mates to be promoted to surgeons. The board itself would consist of five surgeons. In 1828, Congress changed the rank from surgeon's mates to assistant surgeons, which included a time in grade of two years and sea duty before an individual could become eligible before the board.

In February 1837, seven surgeons and twelve others—including Barton and Harris—wrote to President Jackson requesting more surgeons for the fleet. In addition to the request, they wrote it collectively as "the Medical Corps of the U.S. Navy." They had finally become an organized group within the Navy, a professionalized unit with standards and intellectual fervor. Shortly before he left office, Jackson approved the request. Three years later, the Navy included 131 surgeons, passed assistant surgeons, and assistant surgeons. The medical staff corps that had emerged in the Navy directly contributed to its overall organization and professionalism of a more modern officer corps.

Religion and the Chaplain Corps

The United States experienced a revival of religious fervor in the 1830s, which seeped into naval culture and regulations. The issues of temperance in alcohol use, the abolition of slavery, and formalized religious services and institutions within the Navy reflected that broader national experience of the evangelical movement. The decade saw the emergence of Baptists and Methodists founding colleges and theological seminaries and the organization of churches as immigrants began to pour into the country, bringing with them their religious faith and traditions.[59] Cities and towns had, by this decade, become large enough to merit church structures, congregations, and clergy and, with them, regular services. Twice as many Americans were part of an organized religion as in 1776.[60] "The religious awakenings of the early nineteenth century marshalled powerful energies in an age when few other social agencies in the United States had the capacity to do so."[61]

Court-martial records of the era suggest that in many ways Navy sailors were no different from those of any era. When on shore they drank, often to excess, and when they drank, they engaged in activity deemed inappropriate in some cases and illegal in others that ranged from engaging with prostitutes, vocalizing their objections to their superior officers, or committing crimes.

Upon return from the Mediterranean aboard USS *Java* in 1830, Chaplain Hervey Hayes spoke with New York City businessman Arthur Tappan in whom he found a willing ear about the immoral behavior of the sailors and officers of the squadron. Tappan's brother Benjamin had deeply involved himself in the new Democratic Party and served as an elector during the presidential election of 1832.[62] Another brother, Lewis, was an abolitionist. With Lewis and Samuel F. B. Morse, Arthur Tappan founded the *New York Journal of Commerce* and made it free of "immoral advertisements."[63] Tappan later cofounded the American Anti-Slavery Society with William Lloyd Garrison, and Tappan would serve as its president until 1840. After Hayes confided in Tappan, the latter sent a letter to Senator Theodore Frelinghuysen of New Jersey, known as the "Christian Statesman" for his involvement in various religious societies.[64] The letter soon found its way to Levi Woodbury.[65]

Hayes had told a tale of rampant immorality of admitting abandoned females onto the ships for two nights after ships arrived in port. No ship was "free from this pollution."[66] He had said that "it would shock every decent person in the nation who has not been familiar with such scenes to read a true account of the loathsome brutality that disgraces some of our public ships." Hayes omitted no rank from responsibility for the deplorable. He retained high praise for Commodore Rodgers who, he averred, "prohibited the admission of abandoned females into his own ship and forbade their being tolerated in any ships under his command."[67] Rodgers's successor, however, had a far more permissive attitude—at least according to Hayes: "Night after night his ship was swarming with vileness & disease. . . . A Broad Pennant, about twice the ordinary size, proudly fluttered over a squadron of licensed American brothels. . . . We seem to have no shame for ourselves. More men were disabled from duty from [venereal disease] than all others put together on board the ship. For 8 or 9 months together without raising anchor the daughters of pollution like so many Turkey Buzzards were feeding on her carcass."[68]

Those involved in the evangelization of America and the maritime industry sought to rectify that or at least mitigate such actions. Enoch Cobb Wines, who served as a schoolmaster on the frigate USS *Constellation* during its cruise to the Mediterranean in 1829, noted "the moral and religious culture of the seamen on board of our public vessels is too much neglected."[69] Like so many transformative measures, change came not from within the Navy but from beyond its strict confines. While officers on board ships read regular religious services such as funerals, the reason for instituting religious—or rather some semblance of moral guidance—in the Navy initially came from societies particularly through their literature. The Society for Promoting the Gospel among Seamen published its first issue of the *Mariner's Magazine* in 1825. Funded by US and British religious societies, seamen's preachers possessed stations in key ports such as Kronstadt, Le Havre, Rio de Janeiro, and Cape Town.[70]

Formally organized in 1828 in New York City and incorporated in April 1833, the American Seamen's Friend Society intended to "improve the conditions of seamen and bring them into the enjoyment of the same domestic comforts and the same advantages for religious improvement that other men in other walks of life enjoyed."[71] The society would ensure standards for sailors to stay in boardinghouses and would provide reading rooms, savings banks, schools, and frequent prayer meetings as well as the distribution of Bibles. The society's magazine, the *Sailor's Magazine and Naval Journal*, became one of the primary conduits to its members and to a broader national audience. The society's chaplain, the Reverend Charles S. Stewart, became one of the best-known and most prolific authors in the Navy. In the magazine's inaugural issue, Stewart wrote that the whole Navy ought to focus on religion. The magazine itself was to relate the religious experience of pious mariners, the progress of seamen's meetings, churches, and worship at sea as well as the oppression of seamen from government and individuals. It raised awareness of the grievances with Navy hospital funds and advocated for "the efficient introduction of the means of grace on board of our ships of war."[72] Though primarily intended for merchant sailors, sailors on Navy warships were clearly part of this greater effort. In addition to Stewart, who had traveled widely on US Navy ships, including as a missionary to the Sandwich (later Hawaiian) Islands, at least one Navy officer—Lieutenant Benjamin Page—served as a director.

One service offered by the organization was the establishment of register offices where sailors could bring letters from their captain attesting to their good moral character. These registers provided two benefits: First, the sailors would secure recommendations for good boardinghouses, and second, shipmasters would give the bearers first preference when they sought billets at sea. This was the practical aspect of governing and abiding by moral standards. To some, it represented something far more wide reaching.

A writer in the *Sailors Magazine and Naval Journal* using the pseudonym American Spectator opined on the Navy's moral character. The bearers of the national flag to other shores, he wrote, should be regarded as the representatives not only of the heroic spirit but also of the moral virtues of the country.[73] The author noted that the bearing and conduct of American sailors received scrutiny from foreigners. But what were these national traits and virtues? American Spectator was not explicit though one can surmise it was primarily the issue of alcohol use (temperance), based on other articles and efforts in the Navy.

Chaplains suggested requirements for morals included foreswearing cursing, refusing to blaspheme the name of God, and avoiding drunkenness. Alcohol abuse supposedly led to every other problematic issue—disrespectful behavior with superior officers, slothful undertakings on the ship, engagement with prostitutes, theft, and in a few cases murder. "Many of our national ships," American Spectator wrote, "while in foreign ports, instead of carrying with them that healthful moral influence which we might righteously expect, have been made the scenes of the most degrading and brutal vices. . . . The present head of the Navy department has taken decisive steps for an immediate and entire suppression of these evils."[74]

Without some organized effort and dedicated personnel within the Navy, however, any discussion of morals and virtues would have been fruitless. Navy officers took a leadership role in organizations. In June 1833, included among the thirty directors of the American Seamen's Friend Society were six Navy officers, including future commodores Andrew Foote and John D. Sloat. Several were captains. Conspicuous in his absence was one of the most senior officers, Commodore Charles Stewart, likely because his personal life in the 1830s, which included a failed marriage and children with another woman, would not have placed him in the category of the virtuous.[75] To many, the provisioned Bibles and readings from officers were insufficient to provide proper, organized moral guidance. The Navy needed a structural approach. This was achieved in a new Chaplain Corps. "There is indeed an affinity between religion and the sea," wrote one chaplain.[76] It was only natural for military units to reflect their civilian counterparts and have their own spiritual advisors. The latter would provide moral guidance and explain the mysterious forces beyond human control.

The Jacksonian era represented a radical departure from the norm in terms of the professionalization of the Chaplain Corps. The need for spiritual guidance during the broader Age of Sail was of interest for sailors. Only the weather or the actions of their fellow crewmembers would disturb the mysteries of the deep and the monotony of shipboard life. Any interruptions fed sailors' superstitions leading to "lucky days" or "unlucky days," for example.[77] Men still went to sea for months—in the case of whalers it could be years—with no access to family and friends and little news from their town or country. What little they received could be months old. Nevertheless, being receptive to a deity did not necessarily translate into a specific request for religion or religious leaders. For that to occur, other parties had to influence sailors and the Navy. Rather than leave sailors to their own devices, chaplains offered a structured means of explaining events in addition to providing that moral guidance—particularly when they were in port.

Although chaplains had served military units in the United States since 1776, the Navy had no formalized structure for chaplains on warships. Gradually, with the rise of the evangelical movement in the United States and support from nongovernmental organizations such as the American Seamen's Friend Society, the Navy created a professional Chaplain Corps. Prior to the Jacksonian era, religious services in the Navy suffered from a lack of structure and organization. While the Navy authorized individuals to serve as chaplains, it had no selection process, no training pipeline, no formal guidance, and little control over their activity. The only accountability occurred through a court-martial, though this may have only been in theory. In the decade from 1828, the Navy proceeded with 310 courts-martial. In no case was a chaplain prosecuted.[78] The limited number of serving chaplains, the self-controlled role in which they saw themselves as moralizing forces, or a fear from sailors or other officers to make accusations may have been factors. The last possibility is the most unlikely since captains—the highest rank until the Civil War—often found themselves subjects of courts of inquiry and courts-martial.

Secretary Woodbury recognized that the Navy could not realize the moral and religious benefits from having chaplains on ship. This was due to their physical infirmities, their small numbers, and the inadequacy of their pay. In his annual report, Woodbury hoped that Congress would remedy the situation.[79] The need for chaplains increased with the requirement for prayers following Sunday morning muster on ships and "for all naval personnel attached to Navy Yards."[80] The first step in allowing chaplains on board was to grant authority to captains to select their own chaplain. Sometimes they chose the chaplain from the ship's complement. The individuals applying for chaplain positions did not necessarily have to have credentials to be a chaplain; until 1823 it was not necessary that they had served in the position on land or been trained in the ministry. Most simply sought a paid, active-duty commission and would have just as gladly accepted a position as ship's purser or captain's clerk. Some wanted the commission but refused to go to sea.

A chaplain's religious affiliation reflected the times—all were from Christian denominations. Nearly half selected during Jackson's administration were Episcopal with the remainder Presbyterians, Congregationalists, and Baptists. The Navy would not select a Catholic chaplain until 1888 when the long waves of Irish, French Canadian, and other immigrants began to filter into the naval rank and file.[81]

The captain might simply have spoken on behalf of an applicant to the Navy based on their good character, though this practice largely ended after Southard's reforms in 1823, which required all chaplains to be ordained clergymen. The practice of the captain selecting his own chaplain concluded with the appointment of John F. Girard, who served on the frigate *Potomac* from 1840 to 1841.[82] The captain could also, if provided a chaplain, reject him in favor of someone else. From 1778 to 1828, half of the chaplains (forty-three of eighty-five) were captain selectees from the ship's company. Between 1829 and 1840, only one of twelve chaplains received their appointment in that manner.

During any year between 1821 and 1840, the Navy had an average of only nine chaplains.[83] With two-thirds of the fleet deployed at any time, this meant that fewer than half of the Navy's ships had chaplains. This factor was mitigated slightly by the fact that squadrons based in the Mediterranean, for example, might have only one chaplain, but that chaplain could service the several ships based mostly in Port Mahon. However, several chaplains served at Navy yards—such as the Washington Navy Yard, Philadelphia Navy Yard, and the Navy Asylum—further reducing the number available for sea duty. It is also possible, though less likely, that the naval administration believed that fewer chaplain services were required during peacetime than during an armed conflict with its higher casualties and, consequently, funeral services; however, that possibility seems unlikely because chaplains also educated midshipmen while on board. With many ships without a chaplain, an unqualified individual might hold the position only as a collateral duty. This meant that the individual likely had less education than a chaplain would in terms of math, science, literature, history, or languages.

Chaplains during the Jacksonian era were significantly better educated than their predecessors. They were men of letters. In addition to teaching, Navy chaplains wrote some of the most complete perspectives on ship voyages from that era. Chaplain Walter Colton, a Yale graduate, taught moral philosophy at the Scientific and Naval Academy (later Norwich University) before moving to Washington, DC, where he became editor of the *American Spectator* and the *Washington City Chronicle*.[84] He later authored *Deck and Port*, about his Navy voyages in the 1840s. Another Yale graduate, Chaplain George Jones, penned *Sketches of Naval Life with Notices of Men, Manners and Scenery on the Shores of the Mediterranean in a Series of Letters from the* Brandywine *and* Constitution *Frigates* as well as *Excursions to Cairo, Jerusalem, Damascus and Balbec, from the United States Ship* Delaware. Reverend Charles S. Stewart wrote three books on his travels including *A Visit to the South Seas in the U.S. Ship* Vincennes, *during the Years 1829 and 1830.* Stewart would also remain influential during the Jackson years as the editor of the *Naval Magazine*, the first regular publication devoted to the Navy and highly influential for allowing junior officers to propose their ideas for the first time in a mass medium.

Several chaplains also had unique access to presidents, particularly Jackson. Walter Colton preached at a Congregational church attended by Jackson (Jackson would not become a Presbyterian until 1838). Jackson offered Colton, a frequent visitor to the White House, a position either as a consul or Navy chaplain.[85] Colton chose the latter and served as chaplain of the West Indies Squadron beginning in 1831, a curious selection by Jackson since Colton had been a vocal opponent of the president's Indian removal efforts. After one year, Colton proceeded to the *Constitution* for its deployment to the Mediterranean and, later in the decade, served as chaplain of the Philadelphia Naval Station. Much closer to Jackson was William Ryland, a Methodist, who had served as chaplain of the Senate while Jackson served in that body. Ryland served as a chaplain in the Navy during Jackson's administration and became a confidant of the president.[86]

Although chaplains were ostensibly men of God, they were not immune from more secular needs such as pay and rank. Chaplains had no rank, and many felt "materially enfeebled" by the low pay as they received a monthly salary of $40. This amount remained unaltered from 1794 to 1835. Several chaplains appealed directly to Congress in 1832 arguing that while the pay for other officers had increased since the introduction of the Navy, chaplain pay had remained constant. "A chaplain," they suggested, "has no rank except what his pay gives him."[87] But it was to their experience that they based their appeal namely in talent, learning, and piety to command the respect of the other officers. They finally succeeded with "An Act to Regulate the Pay of the Navy of the United States" in which chaplains would receive $1,200 annually on active duty and $800 if on a leave of absence or awaiting orders.[88] Though their pay had substantially increased, it still remained lower than that received by lieutenants.

The 1830s were, therefore, a significant advancement in the creation of a Chaplain Corps. Chaplains were no longer chosen because they were favorites of the captain or educated individuals seeking a job. Instead, they were ministers trained in the faith. They coalesced around the issues of pay and rank thus recognizing that they were a collective group within the rapidly professional naval culture. They reflected their decade, a broader national move toward organized religion and the expectation that people—including sailors—should be held accountable to a higher standard of not just simple laws but rather morals. But even as the nation and the Navy moved to matters of faith, it struggled with the greatest moral issue in its history—enslavement.

Race

No issue divided the young republic more sharply or had a greater lingering impact on the nation than slavery. It permeated nearly every facet of society. The southern economy rested on the cheap labor that the enslaved individuals provided.[89] Naval operations would include enforcement and countering illegal enslavement trade. National discourse during the Jacksonian era was reaching a crescendo that began to manifest into organization and action.

Societal attempts to end enslavement began to appear in newspapers, through literature as well as formal organizations. Oberlin College formally began to recruit free African Americans as students in 1833. That same year, William Lloyd Garrison, Frederick Douglass, and others—including Navy chaplains—founded the American Anti-Slavery Society in Philadelphia. The American Colonization Society, founded two decades before, retained strong political support from such figures as Clay and Jackson and would eventually send over ten thousand previously enslaved individuals to Monrovia in the future state of Liberia. In 1821, Navy lieutenant Robert Field Stockton had negotiated the purchase of land for the Monrovia colony.

Unlike the Army, where African Americans or those of African descent were largely absent until the Civil War (when they were drafted by the necessity of the conflict's scope), African Americans were an integral part of the Navy because of the multicultural merchant fleet from which the Navy often found its sailors.[90] While the precise number of African Americans serving in the Navy is unknown, an article in the *Sailor's Magazine* notes that "the Colored Seamen's Home in New York City, founded in 1839, averaged about 450 boarders a year from the naval and merchant service."[91] Navy sailors were exposed to native cultures and Western European (primarily Spanish and Portuguese) influence in the Gulf of Mexico and South America and to civilizations in Africa, the Middle East, and Asia.

Rear Admiral Charles Steedman recalled his years as a midshipman during the Jacksonian era on his first cruise when he encountered nonwhites in Havana: "The slave-trade was also in full blast and no effort was made to check it; indeed the Governor-General had his coffers filled by shutting his eyes to this horrible traffic."[92] Shortly thereafter at Christmas in San Domingo, Steedman recounted: "None of us felt any hesitation in dancing and flirting with the St. Domingo ladies, although some of them were as black as the ace of spades. Notwithstanding the greater part of them were negroes, I must frankly say that the young women, in grace and lady-like manners, compared most favorably with young women in the best society. . . . We passed some two months at Port au Prince and certainly, although among [them] had a very pleasant and jolly time."[93]

Shipboard corporal punishment was also an indication that the Navy tried to address unjust inequalities ashore. Sailors found guilty of beating a "colored" man received twelve lashes, while kicking a man on the quarterdeck only earned eight lashes. Steedman noted that "beating a colored man" would merit a sailor the same number of lashes as if he had endangered the ship with fire.[94] But like southern society, the Navy sometimes made dispassionate and pragmatic—and most significantly inhumane—decisions about African Americans. Several bases in the South, particularly the important Gosport Naval Shipyard in Portsmouth, Virginia, needed African Americans as laborers. The Board of Navy Commissioners considered their use in 1833. Of every hundred laborers, the board recommended hiring five African Americans. African Americans, the board argued, were less prone to sickness in warm climates and could bear the heat better than their Caucasian counterparts. To support this, they pointed to Navy warships especially in the West Indies, where "the crews of the launches belonging to our ships of war have generally been composed of colored men."[95]

Race relations manifested themselves in the construction of the Norfolk dry dock. The chief engineer of the projects, Loemmi Baldwin Jr., could do little with the cost of granite due to transportation. The only option available to him was to review the labor cost in Norfolk. Specifically, he intended to use local slave labor in the south.[96] Opposition to Baldwin's plans was swift. The president of the Board of Navy Commissioners, Commodore Rodgers, wrote to Baldwin that he had received complaints from Caucasian stonemasons who were being replaced with slave labor. Slaves did not work for free. They earned $0.72 per day and kept $0.10 (the remainder went to their masters), while Caucasians earned $1.50 to $2.[97]

Two years later, the Norfolk commander urged the necessity of employing African American labor as a result of the increase in naval construction. The board temporarily acquiesced but eventually declined the request. The board also authorized Captain Lewis Warrington of the Gosport Navy Yard to employ forty African Americans temporarily and increase their daily rate from sixty-two cents to seventy-five cents, but the Jackson administration denied this. This hiring action also posed a market problem. As noted by

Captain James Barron, then in command of the Norfolk Naval Station, African American laborers near the shipyard and base earned more in the local economy than in the shipyard. Consequently, the shipyard could not obtain a sufficient number. The disparity in pay caused Captain Warrington to state the obvious: Caucasian workers did not perform more labor daily than African Americans; therefore, they should not be entitled to more pay. In addition, he argued, in the extreme climate of the South, African American laborers performed most of the work.[98]

Before the Nat Turner rebellion of 1831, the board suggested that African Americans were not difficult to govern and that they had exhibited no insurrection or disorderly or refractory spirit. Baldwin himself observed on September 27, 1831, that, when he first went to Norfolk, he had strong prejudices against African Americans as laborers and would not employ them. But as he observed all the laborers, he noticed that African Americans worked harder than their Caucasian counterparts. Rodgers and the other two officers of the board advised Woodbury that they could not guarantee employment to only Caucasian laborers because the prohibitive cost of labor would conflict with the public interest. Abandoning the practice of employing African American laborers, they argued, would mean that there would be no way of equalizing wages at different yards. Discharging African American laborers would simply be too expensive.[99] The decision reached by the board, therefore, was not one from an abolitionist standpoint but from the cold, raw numbers associated with construction.

One of the major events that precipitated the American Anti-Slavery Society and abolitionist causes during Jackson's presidency was the Nat Turner slave rebellion. This event, the most violent of the hundreds of slave uprisings in the antebellum period, resulted in the death of fifty-five Caucasians in southeastern Virginia on August 22, 1831. The situation was dire enough that Jackson ordered Marine commandant Archibald Henderson to hold his entire force in readiness for the protection of the headquarters, barracks, arsenals, and other public property that Turner and the rest of the rebelling slaves might attack. In addition, Jackson ordered Henderson to remain silent about the preparations "that no alarm may be created among the Citizens."[100] Captain Jesse Duncan Elliott ordered the USS *Warren* and Marines from two ships to be sent ashore. Turner and his followers were captured and executed, and more than 120 Black people were killed in retaliation.[101]

Following the Turner rebellion, Secretary Woodbury directed a study to assess the use of African American labor over Caucasian artisans.[102] The Board of Naval Commissioners completed its study only a few months later.

The board presented an uncharacteristically lengthy response in their October 15, 1831, letter to Woodbury. In it, the board provided historical background about the subject as well as contemporary arguments and evidence. Despite the publication of the report, Congress found no reason to debate it given the proslavery environment of half of the country. Despite this case, race was less of an issue with the Navy than the nation as a whole. Individuals were held accountable if they attacked an African American because of their race. African Americans and Caucasians served together ashore and at sea. The Navy leadership treated the issue in terms of practicalities rather than of explicit contempt of another race, but the treatment of African Americans by other sailors or officers could be as divergent as the remainder of the country.[103]

Norfolk was not the only site where the enslaved were "leased." The Navy had begun "renting" the enslaved only a few years before in Pensacola. One historian notes that agreements between slaveowners and naval officers were conducted with the expectation that the Navy would be able to "employ" the enslaved, thereby alleviating financial consideration by the slaveowners. The government was under no legal obligation under these "gentlemen agreements" to fulfill payments when construction projects dried up. As every appropriations bill changed, there was no guarantee of the work's permanency, but slavery grew there as a result of the expectation of work on fortifications. Navy agents there "helped slavery secure a foothold in the Gulf Coast."[104]

The first Board of Navy Commissioners in 1817 kept shipyard commandants from employing enslaved African Americans for a variety of reasons: the difficulty of managing those who attempted to escape, the dissent it caused locally among Caucasian laborers who opposed the competition of far cheaper labor, and abuses within the yards themselves.[105] The prohibition remained unchanged until the establishment of the Gosport dry dock, but without adequate reporting or enforcement, the abusive use of slaves by some Navy facilities remained. Racial tensions were also exposed during the Washington Navy Yard strike of 1835, one of the largest users of enslaved labor and free African American labor in the District of Columbia, when strikers fought for reduced workdays. The spark occurred when Commodore Isaac Hull, the yard's commanding officer, forbade workers from entering shops during lunch time or bringing their lunches to work because of concern over thefts. What ensued was the first strike by federal employees, which lasted two weeks. It evolved into a conflict between Caucasian and African American workers, but little was resolved.[106]

Alcohol

Alcohol was abused by sailors, opposed by the temperance movement, and reluctantly condoned by military leadership. In many cases, the use of ardent spirits, beer, and wine was a necessity aboard ships on long cruises, as it had been for centuries of seafarers. The realities of contaminated potable water required an alternative or supplemental drink, whose alcohol content was healthier than some of the water brought aboard. But the reality of spirits in a confined space—or the ample availability in foreign ports after lengthy periods at sea—meant that sailors desperate for physical release from their otherwise mundane duties were more prone to consume more alcohol.

As alcohol abuse grew in the Navy and nationally, a temperance movement emerged that argued for either enforced moderation or simply its prohibition.[107] That, combined with the second great awakening of religion in the making, made for a powerful force. "Making temperance a Christian cause constituted an innovation, for traditional Christianity had not discouraged drinking. . . . Temperance workers paid reformed alcoholics to go on speaking tours, published temperance tracts, put on temperance plays."[108] Not all alcohol was condemned, only distilled liquors. Beer and wine continued to be, if not accepted, at least tolerated. The American Temperance Society was founded in 1826. It was only a matter of time before the movement reached the military.

Alcohol was a perennial problem in both the Army and Navy. Most Navy courts-martial resulted, in part, from the abuse of alcohol leading to dereliction of duty, striking officers, and in rare instances murder. Major General Alexander Macomb, later commander of the ground forces during the Second Seminole War, wrote: "Nothing has tended so much to degrade the rank and file of the Army, as the excessive use of ardent spirit; nor has it been less destructive of their health and discipline."[109]

Alcohol abuse was not limited to military personnel. It was part of a larger societal problem first addressed by Dr. Benjamin Rush, a prominent Philadelphia physician during the revolutionary period. Rush served as surgeon of a Pennsylvania gunboat fleet and was a signer of the Declaration of Independence. In addition, his son Richard was one of the most influential diplomats and civil servants of the Jacksonian era.[110] The senior Rush served as physician general to the military hospitals of the United States and published an *Enquiry into the Spiritous Liquors on the Human Body and Mind* in 1784. His protégé, Dr Edward Cutbush, later published his own work, *Observation on the Means of Preserving the Health of Soldiers and Sailors*, in

1808. Alcohol was an integral part of most Americans' lives. "In 1825, the average American over fifteen years of age consumed seven gallons of alcohol a year."[111] The temperance movement evolved as a Christian cause and during the 1830s reduced the average consumption of alcohol to less than one-third it had been when Jackson became president. It was only natural, therefore, that temperance efforts would affect the military.

On December 28, 1829, the House of Representatives introduced legislation allowing the soldiers and seamen of the United States to forego their daily allotment of alcohol rations. Seamen received six cents in pay for each day they rejected spirits, thus providing a financial inducement to temper alcohol use.[112] In January, Secretary of War John Eaton forwarded communications from Major General Macomb and George Gibson, the commissary general of subsistence, on their views. Eaton himself noted that the daily allowance did not produce a soldier's or sailor's "intemperance" since the quantity was too small.

Instead, he wrote, the fault rested with supplies received from citizens at various posts. The Army tested one plan to "permit a free and unrestrained use of ardent spirits."[113] The theory behind this was that people were less responsible when they confronted more rules. "Mankind," Macomb wrote, "when under too rigid restraint, are more disposed to restiveness, and a violation of rule, than when liberally placed under the guidance of their own sense of propriety."[114] Macomb suggested, instead, that rations of spirits be discontinued and replaced by rice and molasses. The proposed inducement suggested that the sailor or trooper would receive the money accrued at the end of their enlistment contingent upon a certificate from the commanding officer confirming "total abstinence and orderly behaviour."[115] It had occurred previously. In 1820, Secretary of War Calhoun made the same proposal, but participation would be voluntary by post. Where accepted, the inducement was practiced for a time but found to be ineffective in combating alcohol abuse.[116]

The Navy addressed this issue early in Jackson's administration with a House resolution on February 25, 1829, regarding allowing ardent spirit rations to midshipmen. Three Navy medical officers offered testimony to the House of Representatives. Secretary John Branch himself felt it unnecessary to provide his own testimony. Committees asked all to respond on the effect of alcohol on the Navy's morals, health, character, and discipline. Although the three surgeons agreed on the prohibition of ardent spirits, their disparate testimony also cast light on the difference in quality among the Navy's surgeons.

Dr. Lewis Heerman, a German émigré, enlisted as a Navy surgeon in 1801 and, with nearly three decades of experience, could have offered a more analytical response to the congressional inquiry.[117] Instead, his two-page assessment assumed a moralistic tone rather than one of scientific inquiry. The ration of spirits was unnecessary for midshipmen because of their youthful health and vigor, which required no stimulating beverage. Rather, rations created "a perversion of the moral sense" and alcohol was "the bane of our otherwise happy country."[118] The standard half-pint ration was "subversive in the end of rational discipline, and dangerous to the justly proud and chivalric character of the Navy."[119] Heerman continued that alcohol use resulted in "destitution of self-respect, irascibility of temper, ferocity and foolhardiness, as also irrational fear and apprehension of danger, are incidentally enumerated as growing out of intemperance."[120]

The second surgeon's report likewise focused on morality and assumed that a half pint of ardent spirits a day led to inappropriate behavior and a negative impact on health. Its length, sixteen pages of testimony, far surpasses the other reports, but testimony came from individuals, like Heerman, with lengthy service.[121] Enlisting in 1809, William Barton was called upon by Secretary of the Navy Paul Hamilton prior to the War of 1812 to compile regulations for governing naval hospitals and during the war published a lengthy work on the subject.[122] Like Heerman, Barton avoided any analysis except to suggest that rations were not indispensable and that other drinks made from cocoa, chocolate, or coffee would have better effects on the health of naval personnel. Barton called alcohol rations a "poison" that had an insidious and deleterious power on the Navy's youth. He painted a vivid picture of near insane men subject to hallucinations and suicide. The patient's suffering was one of tremors, exhaustion, and agony. Any effort to determine if one could safely determine the quantity of the rations, he argued, would be fruitless. Prone to citing Shakespeare or a poem about Cassio to make his point, he at least suggested that Congress consult the statistical accounts of temperance societies, public charities, and infirmaries.[123]

Surgeon Thomas Harris provided the third report.[124] He had also entered the Navy in 1809 and later oversaw the construction of the Philadelphia Naval Hospital.[125] The character of man, he wrote, is his reason. Alcohol suspended reason and placed the individual under the influence of "an unrestrained or stimulated imagination."[126] Alcohol injured every tissue and organ, and it weakened an individual's constitution that resulted in gout, rheumatism, dropsy, and other ailments. In addition to having no beneficial impact on health, alcohol also played a role in the profession. "An intemperate midshipman will not become a temperate lieutenant or captain."[127] Harris concluded with brief historical analogies that neither the Romans nor "the enthusiastic disciples of the Koran" required ardent rations to found empires. While Heerman and Barton may have repeated the language of the temperance movement, all three were correct in that alcohol could influence behavior in the Navy. In the year leading to their testimony, the Navy had enough examples of how alcohol impacted careers.

Responses to drunken behavior could also be rather draconian as in the case of Matthew Calbraith Perry's father, who was a naval officer during the Revolutionary War. When he found a troublesome midshipman, who had previously attempted to rape a sailor's wife, drunk on watch, he had three of his sailors urinate in the officer's mouth to induce vomiting.[128]

Midshipman Charles G. Hunter of *Java* found himself charged with drunkenness and scandalous conduct after a July 4, 1828, incident. Several officers accused him of being intoxicated and brandishing a cutlass yelling that Commodore John Rodgers was "a damned raskel [*sic*]" and that he would kill Captain Perry if he met him. Among the many charges leveled against him, he was found not guilty of some, guilty of others, and some were not proven for those in which he was guilty; the commodore privately admonished him.[129] A subsequent duel led to Hunter's death, and Jackson ordered Hunter to be erased from the list of officers along with the three other midshipmen involved.[130]

The court-martial list of naval officers abusing alcohol makes a long and depressing list. One sailor found himself charged with habitual drunkenness on the *Porpoise* to the degree where he "could not articulate a syllable."[131] He was found guilty and suspended from the Navy for one year. Marine second lieutenant Francis Neville of the *Hudson* was charged with conduct unbecoming an officer and excessive intoxication as well as seditious and disrespectful conduct to the ship's first lieutenant. He pleaded guilty and was suspended for four months.[132] A sailing master became "flagrantly intoxicated" on the Island of St. Lorenzo and was charged with that act as well as scandalous conduct for violating his oath to abstain from alcohol. He was found guilty and cashiered.[133] A lieutenant aboard the USS *Dolphin* was likewise charged with intoxication and scandalous conduct for repeatedly violating his drinking pledge for which he received a suspension for eighteen months. A midshipman aboard the USS *Guerriere* behaved in a manner that all three surgeons had anticipated with his mental faculties diminished. While in the port of Rio, he became drunk on several occasions and exposed himself "in a shameful manner in the street and public houses . . . to the great disgrace of the character of American officers. Other times he exposed himself in the steerage and on the berth deck to officers and crew."[134]

If the officers of the Navy behaved in such a fashion, so too did the enlisted. But their punishments were considerably more severe. Seaman Nicholas Colby of *Delaware* was drunk at evening quarters and charged with insolence.[135] Others charged and found guilty received extremely disparate punishments based on their actions they took after becoming intoxicated. When Seaman William Hughes of *Delaware* drunkenly struck a petty officer, he was returned to duty without further punishment. But Seaman James Daley of the *Warren* refused an order ashore and told Midshipman George Gay to "go to hell" and received a sentence of two hundred lashes. Commodore Crane mitigated this punishment to one hundred lashes dispensed among the three US ships in port—including the *Delaware*.[136] Seaman Moses Lee on *St. Louis* disobeyed an order by a midshipman while under the influence and received thirty-five lashes.[137]

The issue of ardent spirits raised its head again in 1834. Based on the recommendation of the Board of Navy Commissioners, Secretary of the Navy Woodbury stamped his own imprimatur on how best to deal with ardent spirits. He attempted to revise the laws on naval rations and alcohol rations. Yet, the secretary found himself limited in what he could impose. He wrote: "Finding the component parts of the Navy ration, unlike those of the Army, fixed by law, it was not in my power to diminish or abolish it."[138] By 1842, Congress reduced the ration to one-quarter pint and no one—sailor or officer—could draw more than his ration.

Reorganizing the Marine Corps

In 1830, the Jackson administration proposed dispensing with the Marine Corps. Secretary of the Navy Branch submitted a report to Vice President John Calhoun (serving as president of the Senate) in response to a Senate resolution of March 1, 1830, on the subject.[139] Captain David Conner, commander of the USS *Erie*, had already fitted out his ship without the usual complement of Marines to test the concept of a warship without Marines. Branch refrained from taking a position on the subject, deferring instead to the senior captains in the Navy who were themselves divided on the issue. Branch posed two primary questions to the senior naval leadership: Should Marines be dispensed with, and could seamen assume their duties? With regard to the latter, Branch inquired if older or partially disabled seamen might be used as guards at naval stations. The Marine Corps was modeled on the Royal Marines with duties primarily ensuring ships did not suffer from mutinies by the crew, boarding enemy ships in battle, and defending Navy yards.

The naval leadership was sharply divided. Commodores Daniel Patterson,

Lewis Warrington, William Crane, Isaac Hull, and Charles Ridgely agreed that the Navy could dispense with the Marines. Warrington called the Marines superfluous. Commodore Alexander Dallas went further: "The marine on board . . . is incompetent to the discharge of the seaman's duty. He never evinces the least disposition to acquire a knowledge of the profession and is never seen aloft."[140] Countering arguments that Marines were necessary to deter mutinies, Dallas said that seamen were "willing servants of a free country," and since they were not acquired through impressment like other nations, Marines were unnecessary.

The presiding officer of the Board of Navy Commissioners, Commodore John Rodgers, was the sole officer to take a centrist position, stating that Marines could not be dispensed with, but their total force could be reduced by one-fourth. This would have supported one of the Jacksonian goals to reduce personnel and thus save money.

Presenting the counterargument to retain the Marine Corps were equally experienced commodores including Edmund Kennedy, William Bainbridge, John Orde Creighton, and Charles Stewart. Of all the responses, Stewart's was the lengthiest answer and for good reason. Of those who responded, Stewart had the longest career with commands held during the Quasi-War, Barbary War, War of 1812, as well as with the Mediterranean and Pacific Squadrons. He was junior in the Navy only to James Barron who provided no comments (and who, due to the court results over the Chesapeake-Leopard Affair of 1807, did not serve during the War of 1812). For the final two years of the War of 1812, Stewart commanded USS *Constitution* and had as his senior Marine Lieutenant Archibald Henderson (fig. 6.3), who became commandant of the Marine Corps in 1820. During the *Constitution*'s simultaneous defeat of HMS *Cyane* and HMS *Levant*, Henderson and his Marines provided sufficient fire support for which he later received a medal from Congress for his leadership. Certainly, Henderson was no stranger to Jackson either, having volunteered his services to Jackson during the incursion into Florida in 1818.[141]

Henderson remains the longest-serving commandant of the Marine Corps with nearly thirty years in the position (from 1820 through 1859). Therefore, it is possible that Stewart was in communication with him or at least influenced by him. Retaining Marines, according to Stewart, was proper and necessary.[142] Commenting on the distinguished history of the Marines, with whom he had served during the First Barbary War in particular, he stated that the Marines were the only portion of the crew of a warship that was wholly military. Stewart noted that many people were under the

6.3. Lieutenant Colonel Archibald Henderson. Archibald Henderson led Marines in the Second Seminole War and would serve longer than anyone as Marine Corps commandant. Courtesy Art Collection, National Museum of the Marine Corps, Triangle, Virginia.

impression that most mutinies in the British Navy originated with the practice of impressment. There was, he wrote, "scarcely an instance of the kind on record. Their mutinies originate in oppression, and not in impressment."[143] The Navy was clearly divided about the role of the Marine Corps. It would not be the last time that public officials would question its utility or role. In fact, it would take nearly another century for the real debate to abate as the Marine Corps had proven its expeditionary benefits especially during World War I. For now, however, the fate of the Marine Corps would be on hold for another five years until Jackson attempted reorganization.

The Marine Corps faced the most significant threat to its existence to date at that point when Jackson proposed merging it with the Army. Henderson again led the defense of the organization with the assistance of key Navy captains who directly appealed to Jackson. Jackson, in turn, sent them and Woodbury to Congress to whom the regulations had been referred.[144] Congress and the president agreed that the Marine Corps ought not to be disestablished but reformed. Consequently, Congress passed on June 30, 1834, "An Act for the Better Organization of the Marine Corps," which Jackson signed into law.

The law accomplished several objectives. First, the Marine Corps would remain part of the Department of the Navy. Second, as commandant, Henderson would be promoted to the rank of colonel.[145] Third, the Marine Corps would increase in size by 40 percent.[146] Fourth, enlistment periods would be reduced from five to four years. It also clarified that pay would be equivalent to Army officers and that no Marine could command a Navy yard. The law had an immediate impact on accountability and on the professionalization of the Marine Corps that Henderson had sought throughout his career.

In January 1835, Secretary of the Navy Mahlon Dickerson advised Henderson that all Marines, since the reorganization of the Marine Corps legislation, were subject to the same laws as those governing the Navy.[147] Consequently, the number of naval courts-martial rose dramatically in 1835. From 1829 to 1838, the annual number of courts-martial ranged from ten to thirty-eight. In 1835, there were eighty-two courts-martial.[148] With accountability achieved, and reforms implemented, courts-martial were significantly reduced in subsequent years. Henderson was not in full agreement with the organization, but he obeyed the law. In December 1835, he reported to Dickerson that he considered the Marines ashore and on ships to be detached from his control and that they were now under the command of the local Navy commanders of yards.[149] Unsatisfied with this declarative statement of submission, Henderson again wrote the secretary, expressing his unadulterated views on being under the command of Navy commanders. The Marine Corps had been, he argued, an independent military for nearly forty years with "a spirit of forbearance, deference and good faith."[150] It had worked closely with the Navy. In appealing to the secretary, Henderson invoked the blood that was shed by Marines alongside naval heroes like Captain James Lawrence and aboard other ships like the USS *Constitution*. He reaffirmed his commitment in that he would not question the government but hoped that, in the future, consideration would be made for the Corps' unique military relations.

The 1830s witnessed a confluence of national trends that migrated to the Navy and helped to solidify an emerging naval culture. The religious awakening in the country demanded higher standards of living for the Navy's sailors and officers as well as a sense of national morality that emboldened its sense of maritime destiny over other nations. Scientific curiosity likewise advanced the Navy in incorporating new technology and improving the lives of sailors. With both, a professional, specialized corps of officers reached adolescence and would continue to develop through the nineteenth century. The same was true of the Marine Corps who found their place within the Navy. At the heart of all these advancements were changing expectations for Navy personnel.

Conclusion

The Awakening of the US Navy

The decade of the 1830s under Andrew Jackson's presidency is an important era in the evolution of the US Navy. Evolving navies do not grow at a consistent rate. They advance, stagnate, and sometimes regress in response to a variety of external and internal factors. Externally, key factors affecting a Navy's growth include alliances, conventional and unconventional threats, commercial opportunities, and the availability of rich resources. Internally or domestically, the factors may include the influence of politicians and policy makers, the availability of resources and technologies, the economic ability to build ships or invest in other aspects of a Navy, and the tenor of the naval culture, which either fosters new ideas and visions or suppresses them in favor of more conservative values and traditions. All these factors were evident in the five-decade prelude to Jackson's administration, leading to a period of advancement for the Navy during his terms in office.

From 1815 to 1829, the US Navy tested its strength and logistical ability in international maritime security by deploying its first permanent squadrons to distant stations. The Navy also tested the use of steam power on one ship but found that it lacked the necessary infrastructure to ensure efficiency or expansion across the fleet. Except for the construction of ships of the line, the fleet remained what it had been for several decades. But under Jackson in the 1830s, the Navy evolved physically, intellectually, and socially, experiencing a multifaceted growth spurt that significantly impacted its shape and character well into the late nineteenth century.

During Jackson's presidency, the fleet's size remained constant, but its overall capability improved. The yearslong deployments to distant stations after the War of 1812 had put physical strain on the ships. When Jackson took office, he directed a much-needed reinvestment in the Navy's ships by decommissioning some, repairing others, and building new platforms. In 1831, naval construction spiked as the United States replaced several older and obsolescent ships. Construction of the first dry docks at naval stations also improved the capabilities of the fleet. These dry docks gave the Navy a tool previously afforded only to larger, more mature navies and quietly announced to the world that the Navy was preparing for an enhanced role in global affairs.

During Jackson's tenure, commercial steamships also proliferated, encouraging the transition of steam technology from commercial craft to Navy ships. By the end of the decade, the Navy was building its own steam-propelled frigates.

The 1830s was also a reflective period for officers, who collectively evolved intellectually. Influential officers like Matthew C. Perry fostered junior officers' thinking about naval issues. Establishment of the Naval Lyceum offered junior officers a forum in which to consider and discuss personnel challenges, force structures, strategies, and education. For the first time, many wrote in public forums such as the Lyceum's *Naval Magazine* or other periodicals. Although the junior officers often wrote under pseudonyms to avoid the appearance of open disrespect, they, nonetheless, challenged the conventional thinking of their senior officers. Those senior officers sometimes responded in kind, thus stimulating a vigorous debate for the first time in the Navy's history.

During this period of technological advancements, naval officers began to recognize the inadequacy of their on-the-job education in seamanship and navigation. Naval officers argued for a school to ensure that all new midshipmen learned the standardized science, technology, engineering, and mathematics required to operate and command the expected steam-powered Navy. They nourished the debate on naval education that led to the creation of the Naval Academy.

The Navy also assumed greater responsibility in the world in response to the nation's expanding commercial interests. The Navy had made its initial foray into commerce protection in the West Indies and the Mediterranean around the beginning of the nineteenth century.

Having suppressed the Barbary States in the First and Second Barbary Wars as well as the pirates of the West Indies, the Navy turned its attention to other areas of the globe. New treaties under Jackson meant more trade routes for American merchant ships, which required greater protection by the Navy.

This geographical expansion of patrols meant the Navy increasingly encountered other navies, nations, and cultures. As the 1830s progressed, the Navy was becoming more exploratory and learning about the world. This development was evident at the Naval Lyceum as well as in the fleet itself. Officers collected items from the ancient sites in Greece, native tribes of the Pacific Islands, ancient civilizations of South America, and elsewhere. Officers mailed these artifacts with narratives to the Lyceum, which also served as a museum. Contacts with other navies resulted in meetings and lectures at the Lyceum.[1]

A younger, less evolved Navy would have required a great deal of time and attention from its commander in chief and other leaders. The quickly growing Navy that Jackson inherited, however, had only two primary needs from his perspective: first, accountability for its actions, and second, guidance on what it ought to do. Jackson's approach to leadership coincided with the Navy's needs, and thus, his role for the Navy was a paternal one.

Just as Jackson expected fiscal accountability from the Navy itself, he and senior officers demanded professional accountability from its officers and sailors, demonstrated through adherence to regulations and the good order and discipline of naval personnel. Courts-martial offered the individual accountability necessary to discipline the adolescent officer corps. Proceedings reveal that both officers and sailors were expected to hold to high standards of behavior and that some of them did not. Whereas a sailor's punishment, which might be as severe as a whipping, was administered immediately and then was over, an officer could be discharged and lose a lifetime of pay and prestige. Jackson took direct action in more naval courts-martial than any president from John Adams to James Buchanan (1798–1860). Jackson officially commented on one-quarter of the courts-martial that passed his desk. In one-third of the cases he reviewed, he either overturned the result or granted remission of the punishment recommended by the court-martial board, thereby establishing himself as a firm but compassionate father figure.

Jackson also met the growing Navy's need for guidance and direction. Under his leadership, for the first time, the United States developed a global strategy to respond to crises, act as a deterrent, and protect growing American commerce. The nation was still a third-rate global power, but its quickly expanding economy and merchant fleet needed greater protection. The US fleet numbered only a few dozen ships, but those ships were an important sign to both nonstate actors and nation-states that the country intended to keep a forward presence.

Although Jackson had been an Army general, he recognized the importance of the global maritime environment. His addresses and messages to Congress make clear his belief that America's place in the world was dependent on shipping and that threats posed to that shipping could only be answered by a sufficiently sized Navy. He developed a national maritime strategy, which he expressed both in words (his limited statements) and in deed (the fleet's operational patterns). He promoted economic trade agreements with both traditional and new partners to establish favorable conditions for American merchant ships. He maintained and expanded the squadron system at key geographical locations to protect those ships and ensure security deterrence to potential predators. Finally, he ordered punitive strikes, though only if necessary and as a last resort. He also used the Navy in a supportive role along the coastline of Florida during the Second Seminole War and in riverine operations. Overall, when it came to the high seas, Jackson exercised the judicious use of force.

Jackson's annual messages to Congress, his inaugural addresses, and his farewell address reveal his thoughts on the purpose of a Navy, the pragmatic limitations of the Navy at the time, and its opportunity for growth in defense of the nation and its interests overseas. His executive decisions on diversifying the force structure, incorporating new technologies, expanding global operations in support of commerce, and directing joint Army-Navy activities in the Second Seminole War prove that he was as capable a navalist as the best of his predecessors. Jackson set a clear agenda for the Navy that continued with his successors.

Jackson had a clear understanding of the Navy as a necessary arm of the nation's foreign policy and commercial expansion and of how to project its limited power in the pre-steam era. He had little experience with the Navy before he became president, but he learned quickly and thoroughly. Threats on the high seas and in distant ports underscored the crucial role of the Navy in protecting and promoting his country's interests.

Having found in Jackson a necessary, paternal figure, the Navy needed a purer ideal that transcended politics, justified its existence, and gave it a cause. In the antebellum era, the strongest national rallying cry was "Manifest Destiny"—the desire and expectation that America's boundaries would push west to the Pacific Ocean regardless of what or who lay in its path.[2] Although the term itself was not formally coined until 1845, the impact of the concept was evident during the two prior decades. One notable example of this impact is Jackson's removal of tribes from their ancestral territories as Americans pushed further south and west. Manifest Destiny signaled that a virtuous people were entitled to the land and could seize it by any means necessary. The Army, either through its battles with Indians or in the war with Mexico, was critical in helping the United States actualize Manifest Destiny. The Navy's guiding principle, however, was different.

The Navy was governed by a broader and unlimited concept, that of maritime destiny. Maritime destiny was the belief that America's future lay on the oceans, not merely in its lands. The term, which first appeared in 1834 in

Miriam Coffin by Joseph Hart, predates Manifest Destiny by more than a decade. Its initial appearance in a work of fiction should not diminish its importance or value. On the contrary, its advent suggests the fundamental nature of maritime destiny. Hart's story focused on the Nantucket whaling industry, an enterprise that had been under the protection of the Pacific Squadron since 1817. Hart was part of a generation of early American literary figures who chose the sea as the outlet of their expression. These writers recognized and captured the national spirit of expectancy and hope. That spirit saw US expansion to the Pacific as inevitable but also expected the United States to play a greater role in the world, a role that could be realized only on the sea.

A nation that relies heavily on shipborne trade understands the importance of the sea and is familiar with the tension between the risks and rewards of navigating an unforgiving ocean. The United States, which had relied on shipborne trade since long before its own founding, was no exception. Americans, particularly New Englanders, had also experienced threats by other nations and by pirates. Unsurprisingly then, the United States could appreciate the need for a Navy to protect the wealth that drove her economy and was willing to support it. Unlike the land-locked Army, the Navy was an instrument of the spread of American commerce and, correspondingly, American ideals to foreign shores.

It is not possible to quantify whether maritime destiny was an instrument of naval and national policy or was a perspective that developed and drew the Navy behind it. However, the Navy's operations, the president's vision, and the discourse among officers and literary figures indicate that, by the close of the 1830s, the young Navy had experienced meaningful growth and, in a sense, become self-aware. When Jackson assumed the presidency in 1829, the United States was a third-rate power with a Navy to match. When he left office in 1837, the Navy had professionalized and modernized, grown in capabilities and prestige, broadened its physical and intellectual horizons, and found its purpose. As Tocqueville observed earlier that decade, America was "born to rule the seas." At the close of the 1830s, the Navy was not quite ready to assume that mantle, but it had made strides forward and was acquiring the knowledge and wisdom that would later bring fulfillment of this destiny on wide seas.

Notes

Introduction

1. Robert Remini, *The House: The History of the House of Representatives* (New York: Collins, 2007), 123.

2. Arthur M. Schlesinger Jr., *The Age of Jackson* (Boston: Little, Brown, 1945), 4.

3. Andrew Jackson, *The Papers of Andrew Jackson*, vol. 7, *1829*, ed. Daniel Feller, Harold D. Moser, Laura-Eve Moss, Thomas Coens (Knoxville: University of Tennessee Press, 2007), 15. However, Mark Cheathem in *Old Hickory's Nephew: The Political and Private Struggles of Andrew Jackson Donelson* (Baton Rouge: Louisiana State University Press, 2007) states that the party left on January 19 (54).

4. Carlton Jackson, *Bittersweet Journey: Andrew Jackson's 1829 Inaugural Trip* (Morley, MO: Acclaim Press, 2011) states Jackson "traveled from the Hermitage on the Cumberland, aboard the Steamboat Fair, commanded by Captain Henry Harrison, to Smithland, where the Cumberland joins the Ohio" (35). However, there is no documentation in Carlton Jackson's book to support this claim nor is there evidence in Jackson, *Papers*.

5. S. L. Kotar and J. E. Gessler, *The Steamboat Era: A History of Fulton's Folly on American Rivers, 1807–1860* (Jefferson, NC: McFarland, 2009), 26. Another source is William C. Davis, *The Greatest Fury: The Battle of New Orleans and the Rebirth of America* (New York: Dutton, 2019).

6. Daniel Walker Howe, *What Hath God Wrought: The Transformation of America, 1815–1848* (Oxford: Oxford University Press, 2007), 40.

7. Howe, *What Hath God Wrought*, 214.

8. Kotar and Gessler, *Steamboat Era*, 90.

9. Sources dispute Andrew Jackson Donelson's role during the first Seminole War. For example, Mark Cheathem's *Historical Dictionary of the Jacksonian Era and Manifest Destiny* (Lanham, MD: Rowman and Littlefield, 2017), 125, as well as Mark Crawford, David Stephen Heidler, and Jeanne Heidler's *Encyclopedia of the Mexican-American War* (Santa Barbara: ABC-CLIO, 1999), 100, both clearly state that Donelson served as Jackson's aide during the first Seminole War. He then attended West Point and returned to Florida as Jackson's aide during his role as governor.

10. Pauline Wilcox Burke, *Emily Donelson of Tennessee* (Knoxville: University of Tennessee Press, 2001), 123.

11. Some of the passenger names can be found in Jackson, *Bittersweet Journey*, 50.

12. Jackson, *Bittersweet Journey*, 83.

13. Simon Schama, *The Embarrassment of Riches: An Interpretation of Dutch Culture in the Golden Age* (New York: Alfred A. Knopf, 1988), 379.

14. Frances Trollope, *Domestic Manners of the Americans* (London: Whittaker, Teacher, 1832), see chapter 13 as no page numbers are provided. Online text available from http://www.gutenberg.org.

15. Harold Sprout and Margaret Sprout, *The Rise of American Naval Power, 1776–1918* (Annapolis, MD: Naval Institute Press, 1990), 114.

16. Claude Berube, "Budget Battles, Interest Groups, and Relevancy in a New Era: The Ship of the Line USS *Pennsylvania*," Naval Institute *Proceedings*, January 2008, 62–68.

17. Jackson, *Bittersweet Journey*, 110.

18. Andrew A. Robichaud, *Animal City: The Domestication of America* (Cambridge, MA: Harvard University Press, 2019), 301n5, cites several sources that estimate the crowd size to be between ten thousand and twenty thousand.

19. Mark Leepson, *What So Proudly We Hail: Francis Scott Key, A Life* (New York: St. Martin's Press, 2014), 113.

20. James Richardson, *A Compilation of the Messages and Papers of the Presidents*, vol. 2, *1817–1833*; Andrew Jackson, First Inaugural Address, March 4, 1829, http://avalon.law.yale.edu/19th_century/jackson1.asp.

21. Richardson, *Compilation of the Messages and Papers*, 2:438.

22. Mark R. Cheathem ("The Stubborn Mythology of Andrew Jackson," *Reviews in American History* 47 no. 3 [September 2019]: 342–48) reviews recent literature on Jackson and dispels several apocryphal stories such as Rachel Jackson suffering from a heart attack as a result of how the media treated her during the 1828 campaign or that Jackson adopted Lyncoya, a Native American child.

23. The "brief" history can be considered as either fifty years or thirty years. In the case of the former, the Second Continental Congress built a Navy during the American Revolution in the 1770s, but the Navy was disbanded after the war. A better argument is the case of the latter since the Department of the Navy was not established until 1798.

Chapter 1

1. Craig Symonds, *Navalists and Antinavalists: The Naval Policy Debate in the U.S., 1785–1827* (Wilmington: University of Delaware Press, 1980).

2. Augustus Granville Stapleton, *George Canning and His Times* (London: John W. Parker and Son, 1859), 394–96; Harold Temperley, *The Foreign Policy of Canning, 1822–27* (Abington, UK: Frank Cass, 1925), 127

3. Gayle Penmer, "Historic Notes and Documents: The Mix Diaries," *Florida Historical Society* 82, no. 2 (Fall 2003): 191–218.

4. A few early twenty-first-century Jackson biographies include H. W. Brands, *Andrew Jackson: His Life and Times* (New York: Random House, 2005); Jon Meacham, *American Lion: Andrew Jackson in the White House* (New York: Random House, 2008); Mark Cheathem, *Andrew Jackson: Southerner* (Baton Rouge: Louisiana State University Press, 2013); Andrew Burstein, *The Passions of Andrew Jackson* (New York: Random House, 2003).

5. Alexis de Tocqueville, *Democracy in America* (New York: Harper and Row, 1965), 22.

6. William MacDonald, *Jacksonian Democracy, 1829–1837*, vol. 15 of *The American Nation: A History* (New York: Harper Brothers, 1907), 26

7. David Wootton, ed., *The Essential Federalist and Anti-Federalist Papers* (Indianapolis: Hackett, 2003).

8. Samuel J. Watson, *Jackson's Sword: The Army Officer Corps on the American Frontier, 1810–1821* (Lawrence: University Press of Kansas, 2012), 20.

9. Tocqueville, *Democracy in America*, 350.

10. Williamson Murray, *The Shaping of Grand Strategy: Policy, Diplomacy and War* (Cambridge: Cambridge University Press, 2011), 11.

11. Robert P. Wettemann Jr., *Privilege vs. Equality: Civil-Military Relations in the Jacksonian Era, 1815–1845* (Santa Barbara, CA: ABC-CLIO, 2009), 20.

12. Ian Toll, *Six Frigates: The Epic History of the U.S. Navy* (New York: W. W. Norton, 2006), 58–61.

13. David F. Long, *Gold Braid and Foreign Relations: Diplomatic Activities of U.S. Naval Officers, 1798–1883* (Annapolis, MD: Naval Institute Press, 1988), 13.

14. Claude Berube and John Rodgaard, *A Call to the Sea: Captain Charles Stewart of the USS* Constitution (Washington, DC: Potomac Books, 2005), 187–88.

15. For Porter and the Fajardo Affair, see Michael Birkner, "The Foxardo Affair Revisited: Porter, Pirates, and the Problem of Civilian Authority in the Early Republic," in Watson, *Warfare in the USA*, 165–78. For Downes and the Quallah Battoo attack, see John M. Belohlavek, "Andrew Jackson and the Malaysian Pirates: A Question of Diplomacy and Politics," *Tennessee Historical Quarterly* 36 no. 1 (Spring 1977): 19–29.

16. Christopher McKee, *Ungentle Goodnights* (Annapolis, MD: Naval Institute Press, 2018), 142.

17. Berube and Rodgaard, *Call to the Sea*, 7–10.

18. No former naval officer became president until the election of 1960. Both party nominees, John Kennedy and Richard Nixon, had served in the Navy during World War II.

19. Matthew Warshauer, *Andrew Jackson and the Politics of Martial Law: Nationalism, Civil Liberties, and Partisanship* (Knoxville: University of Tennessee Press, 2006), 19–45.

20. Wettemann, *Privilege vs. Equality*, 68.

21. Henry Clay Jr. to Henry Clay, September 16, 1830, Henry Clay, *The Papers of Henry Clay*, vol. 8, *Candidate, Compromiser, Whig, March 5, 1829–December 31, 1836*, ed. Robert Seager II and Melba Porter Hay (Lexington: University of Kentucky Press, 1984), 265.

22. Wettemann, *Privilege vs. Equality*, 68–69.

23. Robert Remini, interview with the author in the House of Representatives, Office of the Historian, June 24, 2009.

24. The status of the Navy in 1828 can be found in the "Annual Report of the Secretary of the Navy," in *American State Papers*, vol. 3, *Naval Affairs*, November 27, 1828, delivered to Congress on December 2, 1828 (Washington, DC: Gales and Seaton, 1861).

25. Samuel Southard to John Quincy Adams February 9, 1829, US Navy Department, Letters Sent by the Secretary of the Navy to the President and Executive Agencies, National Archives and Records Administration, Washington, DC, archives.gov (hereafter NARA).

26. "Annual Report of the Secretary of the Navy," 1828, in *American State Papers*.

27. *American State Papers*, vol. 3, 1860213.

28. US Congress, "Annual Report of the Secretary of the Navy," 1828, in *American State Papers*.

29. US Congress, "Annual Report of the Secretary of the Navy," 1828, in *American State Papers*.

30. Paul Nagel, *John Quincy Adams: A Public Life, a Private Life* (New York: Alfred P. Knopf, 1997); Fred Kaplan, *John Quincy Adams: American Visionary* (New York: Harper, 2014).

31. John Quincy Adams, *Memoirs of John Quincy Adams: Comprising Portions of His Diary from 1795–1848*, vol. 5, ed. Charles Francis Adams (Philadelphia: Lipincott, 1875), 136.

32. Charles N. Edel, *Nation Builder: John Quincy Adams and the Grand Strategy of the Republic* (Cambridge, MA: Harvard University Press, 2014), 97.

33. Edel, *Nation Builder*, 129

34. David Foster Long, *Sailor-Diplomat: A Biography of Commodore James Biddle, 1783–1848* (Boston: Northeastern University Press, 2007), 156.

35. Burstein, *Passions of Andrew Jackson*, 131.

36. Edel, *Nation Builder*, 149.

37. May 5, 1818, as cited in Burstein, *Passions of Andrew Jackson*, 131.

38. David S. Heidler and Jeanne T. Heidler, *Old Hickory's War: Andrew Jackson and the Quest for Empire* (Mechanicsburg, PA: Stackpole Books, 1996), 203–18.

39. Edel, *Nation Builder*, 227.

40. Edel, *Nation Builder*, 237.

Chapter 2

1. Paul Starr, *The Creation of the Media: Political Origins of Modern Communication* (New York: Basic Books, 2004), 124.

2. Carl Kaestle, Helen Damon-Moore, Lawrence Stedman, and Katherine Tinsley, *Literacy Rates in the United States* (New Haven, CT: Yale University Press, 1991), 23.

3. The records of the Naval Lyceum are held at the US Naval Academy Museum.

4. N. A. M. Rodger, "Navies and the Enlightenment," *Essays in Naval History* (Surrey: Ashgate, 2009), xiv.

5. Claude G. Berube, "Born to Rule the Seas: The Navy during Andrew Jackson's Presidency and the Genesis of American Naval Power," PhD diss., University of Leeds, 2018.

6. Edgar Allan Poe, "Critical Notices: Maury's Navigation," *Southern Literary Messenger* 2.7 (1836): 454–55.

7. Andrew Taylor and Aine Kelly, eds., *The Idea of America* (New York: Routledge, 2013), 100.

8. Tim Lanzendörfer, "Naval Biography, The War of 1812, and the Contestation of American National Identity," in *Warring for America: Cultural Contests in the Era of 1812*, ed. Nicole Eustace and Frederika Teute (Chapel Hill: University of North Carolina Press, 2017), 289–91.

9. Samuel F. Du Pont to Charles Davis, April 23, 1835, Samuel F. Du Pont Papers, Hagley Museum and Library, Brandywine, DE (hereafter Du Pont Papers). The Du Pont collection provides the richest sole source of junior officer correspondence during this period.

10. Garrett J. Pendergrast to Samuel F. Du Pont, January 5, 1824, Louisville, KY, Du Pont Papers, Group 9. Pendergrast was the son-in-law of Commodore James Barron, the senior-most officer in the Navy until his death. Pendergrast also commanded USS *Cumberland* at the outset of the Civil War.

11. Benjamin F. Sands, *From Reefer to Rear-Admiral: Reminiscences and Journal Jottings of Nearly Half a Century of Naval Life* (New York: Frederick A. Stokes, 1899), 11.

12. Daniel H. Mackay to Samuel F. Du Pont, May 26, 1829, Du Pont Papers.

13. Records of General Courts-Martial and Courts of Inquiry of the Navy Department, 1799–1867, Navy Courts-Martial, 1825–1840, Case 480, NARA.

14. Records of General Courts-Martial, Case 480.

15. Daniel Mackay to Samuel Du Pont, May 26, 1829, Du Pont Papers.

16. John S. Wiley to Samuel Du Pont, January 23, 1831, USS *Ontario*, Du Pont Papers.

17. Thomas Lambert to John Hale, August 26, 1835, Thomas Lambert Papers, New Hampshire Historical Society, Concord.

18. Thomas Lambert to John Hale, July 30, 1834, Thomas Lambert Papers, New Hampshire Historical Society, Concord.

19. Daniel H. Mackay to Samuel F. Du Pont, Philadelphia, December 5, 1828, Du Pont Papers.

20. Garrett A. Pendegrast to Samuel F. Du Pont, [n.d.], Du Pont Papers.

21. Daniel H. Mackay to Samuel F. Du Pont, August 29, 1829, Du Pont Papers.

22. Daniel H. Mackay to Samuel F. Du Pont, January 1, 1829, Du Pont Papers.

23. Thomas Lambert to John Hale USS *Brandywine* 20 May 1834; Thomas Lambert Papers, New Hampshire Historical Society.

24. Arnold [Mandain?] to Samuel F. Du Pont, February 21, 1835, Du Pont Papers.

25. Samuel F. Du Pont to John S. Wiley, November 22, 1832, Du Pont Papers.

26. George W. Rodgers, Order of March 16, 1832, Letter Books of US Naval Officers, March 1778–July 1908, Letters sent by Captain George W. Rodgers, commanding the *Warren*, December1831–May 1832, NARA.

27. Charles Steedman, *Memoir and Correspondence, Rear Admiral, United States Navy, with His Autobiography and Private Journals, 1811–1890* (Cambridge, MA: Riverside Press, 1912), 25.

28. Unknown author, "To the Officers," *Military and Naval Magazine* 4 (September 1834): 177.

29. Peter Simple (pseudonym), "Naval Sketches," *Military and Naval Magazine* 1 (March–August 1833): 376.

30. Coquille, "To the Officers of the Two Services," *Naval and Military Magazine* 1 (March–August 1833): 303–4.

31. Benjamin Homans, *Army and Navy Chronicle*, January 3, 1835, 1.

32. *Sailor's Magazine and Naval Journal* 1 (August 1829): iii.

33. Reverend Charles S. Stewart, editor, *Naval Magazine*, January 1836, 4.

34. John H. Schroeder, *Matthew Calbraith Perry: Antebellum Sailor and Diplomat* (Annapolis, MD: Naval Institute Press, 2001), 72.

35. Sinclair, "The Navy," *Naval Magazine*, October 1834, 113–14.

36. Untitled article, *Baltimore American*, reprinted in *Army and Navy Chronicle* 2, no. 14 (April 1836): 223.

37. D, "Thoughts on the Present Condition of the Navy, and Suggestions for Its Improvement," *Military and Naval Magazine*, January 1834, 263.

38. Charles Dupin, "Naval Force of Great Britain," reprinted in the *Military and Naval Magazine*, September 1835, 28–32.

39. A Lieutenant in the Navy, "Guyaquil," *Naval Magazine*, January 1836, 227–35.

40. "Customs and Traditions, Navy," Naval History and Heritage Command, https://www.history.Navy.mil, accessed December 8, 2020.

41. A Junior Officer in the Navy, "Petty Officers of the Navy," *Military and Naval Magazine*, September 1833 to February 1834, 114.

42. An Officer, "Naval Education," *Naval Magazine*, May 1836, 219.

43. Editorial, "On Naval Schools," reprinted in *Army and Navy Chronicle*, July 18, 1839, 44.

44. A.S., "Thoughts on the Navy," *Naval Magazine*, January 1837, 22.

45. Data compiled by the author based on reports of the secretary of the Navy and Edward

Callahan, ed., *List of Officers of the Navy of the United States and Marine Corps from 1775 to 1900* (New York: L. R. Hamersley, 1901), appendix B2 and B3.

46. D., "Thoughts on the Present Conditions," 257–65.

47. Sinclair, "The Navy," 114.

48. C., "Admirals," *Army and Navy Chronicle*, January 4, 1838, 12. The author was likely Commodore Charles Stewart, the second-longest serving officer in 1838, who commanded *Pennsylvania* out of the shipyard the year before.

49. Anonymous letter, *Army and Navy Chronicle*, September 12, 1839, 168.

50. Robert D. Madison, "Cooper's Place in American Naval Writing," presented at the Fourth Cooper Seminar, James Fenimore Cooper: His Country and His Art, State University of New York College at Oneonta, July 1982, external.oneonta.edu/cooper/articles/suny/1982suny-madison1.html, accessed December 29, 2017.

51. James Kirke Paulding to Samuel Southard, May 5, 1828, James Kirke Paulding, *The Letters of James Kirke Paulding*, ed. Ralph M. Aderman (Madison: University of Wisconsin Press, 1962), 95–97.

52. Jeffrey H. Hacker, ed., *The Early Republic and Rise of National Identity, 1783–1861* (New York: Routledge, 2014), 85.

53. Perry organized a dinner in New York City in 1824 for his fellow officers and Cooper.

54. Washington Irving to Peter Irving, December 30, 1812, Washington Irving, *Complete Works of Washington Irving: Letters 1802–1833*, ed. Ralph M. Aderman, H. L. Kleinfield, and Jenifer S. Banks (Boston: Twayne, 1978), 350–51.

55. Washington Irving to Alexander Hill Everett, July 22, 1829, Irving, *The Complete Works of Washington Irving*, vol. 1, *1803–1859*, ed. Wayne R. Kime (Boston: Twayne, 1981), 449.

56. Irving to Everett, Irving, *Complete Works*, 1:449.

57. Washington Irving to Peter Irving, December 3, 1830, Irving, *Complete Works*, 1:565.

58. Washington Irving to Martin Van Buren, April 30, 1838, Library of Congress. See also S. T. Williams, *The Life of Washington Irving* (Oxford: Oxford University Press, 1935), 67.

59. James Kirke Paulding to Samuel Southard, May 5, 1828, Paulding, *Letters*, 95–97. Some of the recommended works were: "[Nathaniel] Bowditch's Naval Gunnery, [Sir Howard] Douglas on Marine Artillery, [James] Marshall on the Equipment of Guns [A Description of Commander Marshall's New Mode of Mounting and Working Ship's Guns], [John] Charnock's Naval Architecture, [George] Vancouver's Voyage [A Voyage of Discovery to the North Pacific Ocean], Analectic Magazine, [Francois Michaux] North American Sylva, On the Preservation of Timber."

60. James Kirke Paulding to Martin Van Buren, December 6, 1834, Paulding, *Letters*, 153–55.

61. James Kirke Paulding to Washington Irving, July 30, 1838, Paulding, *Letters*, 220–22.

62. James Kirke Paulding to James Fenimore Cooper, May 20, 1839, Paulding, *Letters*, 256–57.

63. By virtue of Goldsborough's position, he was able to secure a commission for his son Louis M. Goldsborough, who became the youngest midshipman at the age of seven but only reported for duty when he was eleven.

64. Madison, "Cooper's Place in American Naval Writing."

65. Pennsylvanian, "Book Review," *Army and Navy Chronicle*, May 16, 1839, 317.

66. Anonymous, no article title, *United Service Gazette*, June 1839, 37.

67. Tripoli Monument Accession Folder, US Naval Academy Museum, Annapolis, MD.

68. Robert Gross and Mary Kelley, eds., *An Extensive Republic: Print, Culture and Society in the New Nation, 1790–1840* (Chapel Hill: University of North Carolina Press, 2010), 305.

69. Carl F. Kaestle, *Pillars of the Republic: Common Schools and American Society, 1780–1860* (New York: Hill and Wang, 1983), 63.

70. See Mark Groen, "The Whig Party and the Rise of Common Schools, 1837–1854," *American Educational History Journal* 35, iss. 1/2 (Spring/Summer 2008): 251–60. Mann's own seminal speech at Brown University, "The Progressive Character of the Human Race" was, unfortunately, never preserved.

71. Kaestle, *Pillars of the Republic*, 99.

72. Joseph Churchman and James Churchman, *Rudiments of National Knowledge, Presented to the Youth of the United States, and to Enquiring Foreigners by a Citizen of Pennsylvania* (Philadelphia: E. L. Cary and A. Hart, 1833).

73. H. W. Dickinson, *Educating the Royal Navy: Eighteenth- and Nineteenth-Century Education for Officers* (London: Routledge, 2007), 10.

74. Dickinson, *Educating the Royal Navy*, 29.

75. A review of this topic can be found in David McLean, *Education and Empire: Naval Tradition and England's Elite Schooling* (London: British American Press, 1999).

76. Dickinson, *Educating the Royal Navy*, 32.

77. Captain Charles Stewart to the Secretary of the Navy, November 26, 1814, in *American State Papers*, vol. 4 (Washington: Gale and Seaton, 1834), 356.

78. Stephen Ambrose, *Duty, Honor, Country. A History of West Point* (Baltimore: Johns Hopkins University Press, 1966), 22.

79. The Navy book lists can be found in a variety of sources and in Record Group 45, the Navy Records, NARA, or in secondary sources such as Hester Blum, *The View from the Masthead: Maritime Imagination and Antebellum American Sea Narratives* (Chapel Hill: University of North Carolina Press, 2008), 36.

80. Journals of Thomas A. Dornin, US Navy, (1826–55), diary, microfilm publication M981, Roll #1, August 4, 1831, NARA.

81. Steedman, *Memoir and Correspondence*, 121.

82. Duer served as president of Columbia from 1829 to 1842. William Duer to Isaac Chauncey, February 3, 1830, letter found in House Committee on Naval Affairs, HR21A-D17.5, Center for Legislative Archives, NARA.

83. Isaac Chauncey to William Duer, March 3, 1830, February 3, 1830, letters found in House Committee on Naval Affairs, HR21A-D17.5, Center for Legislative Archives, NARA.

84. This information is based on email communications of June 1, 2011, between the author and the Columbia University archivist. By her admission, Columbia has minimal correspondence from the early nineteenth century. On September 3, 1833, Edward Livingston aboard the *Delaware* wrote to Duer thanking him for his diploma, although Livingston is not listed among Columbia alumni, nor is he in the 1830 to 1833 matriculation book, nor was he the recipient of an honorary degree.

85. Christopher McKee, *A Gentlemanly and Honorable Profession* (Annapolis, MD: Naval Institute Press, 1991), 463–44.

86. Angela G. Ray, *The Lyceum and Public Culture in the Nineteenth Century United States* (East Lansing: Michigan State University Press, 2005.)

87. Preble, "The Navy," *Military and Naval Magazine*, November 1834, 163

88. William P. Leeman, *The Long Road to Annapolis: The Founding of the Naval Academy and*

the Emerging American Republic (Chapel Hill: University of North Carolina Press, 2010), 48, cited in Eugene S. Ferguson, *Truxtun of the Constellation: The Life of Commodore Thomas Truxtun, US Navy, 1755–1822* (Baltimore: Johns Hopkins University Press, 1956), 128.

89. Leeman, *Long Road*, 84.

90. Congressman Robert Hayne, US Congress Register of Debates, cited in Leeman, *Long Road*, 90.

91. A.B., "The Navy," *Military and Naval Magazine*, April 1835, 83.

92. Boston, New York, Philadelphia, Norfolk were the four primary homeports. To a lesser degree Portsmouth New Hampshire, Baltimore, and Washington, DC, serviced at least one ship for one or more years. Pensacola was established later in the decade.

93. "United States Military Philosophical Society," *National Intelligencer and Washington Advertiser*, August 4, 1806, 2.

94. Joseph Kett, *The Pursuit of Knowledge under Difficulties: From Self-Improvement to Adult Education in America, 1750–1790* (Stanford, CA: Stanford University Press, 1994), xv.

95. Uncredited, "U.S. Naval Lyceum," *Naval Magazine*, January 1836, 5.

96. Letter from M. C. Perry, C. O. Handy, and J. Haslett to Board of Naval Commissioners, December 10, 1833, Letterbook of the U.S. Naval Lyceum, 1834–1871, Box 37, US Naval Academy Museum.

97. Writer, "To the Officers of the Army and Navy," *Military and Naval Magazine*, September 1834 to February 1835, 177

98. This event is captured in a painting by William Edward West (1788–1857).

99. Schroeder, *Matthew Calbraith Perry*, 70.

100. Andrew Jackson was an honorary member of the Columbian Institute.

101. All figures on the names and ranks of members belonging to the Naval Lyceum were compiled by the author in "Born to Rule the Sea" but not included in the index.

102. Henry Willett to Jesse D. Elliott, March 13, 1835, Naval Lyceum Records, US Naval Academy Museum. The Lyceum offered the use of rooms, the library, and papers during the visit of the USS *Constitution* and the USS *Peacock*.

103. Lists compiled from Record Book: List of Visitors, U.S. Navy Lyceum, 1834–1843, Box 24, Naval Lyceum Records, US Naval Academy Museum.

104. Joseph C. Hart, *Miriam Coffin—or, The Whale Fisherman* (New York: Harper, 1835), 196.

Chapter 3

1. The concept of the navalists and antinavalists is derived from Symonds, *Navalists and Anti-Navalists*.

2. Berube, "Budget Battles, Interest Groups, and Relevancy," 62–68.

3. US Department of Commerce, "Documented Merchant Vessels—Composition of the Merchant Marine: 1789 to 1945," in *Historical Statistics of the United States*, series K 94–104 (Washington, DC: Government Printing Office, 1949), 207.

4. US Department of Commerce, "Net Tonnage Capacity of Vessels Entered and Cleared: 1789 to 1970," in *Historical Statistics of the United States*, series Q 506–17, 760.

5. Malina Johar Schueller, *U.S. Orientalisms: Race, Nation, Gender and Literature, 1780–1890* (Ann Arbor: University of Michigan Press, 1998), 27.

6. Richardson, *Compilation of the Messages and Papers*, 2:438.

7. Jackson, First Inaugural Address, March 4, 1829.

8. Richard P. Longaker, "Was Jackson's Kitchen Cabinet a Cabinet," *Mississippi Valley Historical Review* 4, no. 1 (June 1957): 94–108.

9. Robert Remini, interview with the author, Washington DC, House of Representatives, June 24, 2009.

10. I have changed "Peggy" to "Margaret" Eaton since John Marszalek makes clear in his study of the Eaton affair that neither she nor her family called her "Peggy." John Marszalek, *The Petticoat Affair: Manners, Mutiny, and Sex in Andrew Jackson's White House* (Baton Rouge: Louisiana State University Press, 2000), 4.

11. Meacham, *American Lion*, 67.

12. John M. Belohlavek, *Andrew Jackson: Principle and Prejudice* (New York: Routledge, 2016), 65.

13. Robert Watson, *Affairs of State: The Untold History of Love, Sex, and Scandal, 1789–1900* (New York: Routledge, 2012), 193.

14. S. 111, April 11, 1820, Bills and Resolutions of the US Senate, 16th Cong., 1st Sess.

15. Marszalek, *Petticoat Affair*, 40.

16. Burstein, *Passions of Andrew Jackson*, 178; Mark Cheathem, "The High Minded Honourable Man: Honor, Kindship, and Conflict in the Life of Andrew Jackson Donelson," *Journal of the Early Republic* 27 (Summer 2007): 273–77.

17. Andrew Jackson to Mary Ann Eastin, October 24, 1830, Andrew Jackson, *The Papers of Andrew Jackson*, vol. 3, *1814–1815*, ed. Harold D. Moser, David R. Hoth, Sharon Macpherson, and John H. Reinbold (Knoxville: University of Tennessee Press, 1991), 579.

18. Andrew Jackson to Andrew Jackson Donelson, April 19, 1831, Jackson, *Papers*, 3:192

19. John Coffee to Andrew Jackson, May 13, 1831, Andrew Jackson, *The Papers of Andrew Jackson*, vol. 9, *1831*, ed. Daniel Feller, Laura-Eve Moss, Thomas Coens, Erik B. Alexander (Knoxville: University of Tennessee Press, 2013), 244.

20. Andrew Jackson Donelson to Andrew Jackson, June 19, 1831, Jackson, *Papers*, 9:315.

21. Message of the President of the United States to Both Houses of Congress, December 7, 1830, *Register of Debates in Congress*, vol. 7 (Washington, DC: Gales and Seaton, 1831), appendix, xii.

22. Message of the President, appendix, xii.

23. For information on the society, see Sidney Forman, "The United States Military Philosophical Society, 1802–1813," *William and Mary Quarterly* 3rd ser., vol. 2, no. 3 (July 1945): 273–85.

24. Message of the President, appendix, 2.

25. Message of the President, appendix, xii.

26. Berube, "Born to Rule the Seas,", fig. 3.1, 114, and fig. 3.2, 115, compiled and calculated from Annual Reports of the Secretary of the Navy.

27. John H. Schroeder, "Jacksonian Naval Policy 1829–37," in *New Aspects of Naval History: Selected Papers from the 5th Naval History Symposium*, ed. by the Department of History, US Naval Academy (Baltimore: Nautical and Aviation Publishing Company of America, 1985), 122.

28. Amos Kendall to his wife, March 27, 1829, Amos Kendall, *Autobiography of Amos Kendall*, ed. William Stickney (Boston: Lee and Shepherd, 1872), 314.

29. "Annual Report of the Secretary of the Navy," in *American State Papers*, 3:351.

30. "Annual Report of the Secretary of the Navy," in *American State Papers*, vol. 4, *Naval Affairs* (Washington, DC: Gales and Seaton, 1861), 590.

31. Record Group 45, Records of Boards and Commissions, index to the records, NARA.

32. One of the significant assessments of the board was David F. Long's chapter, "The Navy under the Board of Naval Commissioners," in *In Peace and War: Interpretations of Naval History*, ed. Kenneth Hagan (Westport, CT: Greenwood Press, 1984), 63–64.

33. Richard Rathbun, "The Columbian Institute for the Promotion of Arts and Sciences," Smithsonian Institution bulletin (Washington, DC: Government Printing Office, 1912), 19–20. Jackson was a resident member, but there is no indication that he ever attended a meeting or encountered Rodgers in this setting.

34. John Rodgers to John Branch, November 3, 1829, in *American State Papers*, 3:396–402.

35. John Rodgers to John Branch, November 3, 1829, in *American State Papers*, 3:396–402.

36. John Branch to Board of Navy Commissioners, November 13, 1829, in *American State Papers*, 3:396.

37. Albert Somit, "Andrew Jackson as Administrator," *Public Administration Review* vol. 8, no. 3 (Summer 1948): 190.

38. John Branch to Andrew Jackson, October 17, 1829, US Navy Department, Letters Sent by the Secretary of the Navy to the President and Executive Agencies, 1821–1886, Letters Sent by the Secretary of the Navy to Officers, 1798–1871, microfilm VB255.1.U72, Roll 1, July 21, 1821–September 30, 1836, NARA.

39. John Rodgers to John Branch, October 27, 1829, in *American State Papers*, 3:396–402.

40. Sprout and Sprout, *Rise of American Naval Power*, 114.

41. The Navy Department to Michael Hoffman, *New American State Papers*, vol. 6, *Administration*, ed. by K. Jack Bauer (Washington, DC: Scholarly Resources Inc, 1981), 18.

42. From 1842 to 1862, this system included: Bureau of Naval Yards and Docks; Bureau of Construction, Equipment, and Repairs; Bureau of Provisions and Clothing; Bureau of Ordnance and Hydrography; and Bureau of Medicine and Surgery.

43. Donald B. Cole, *A Jackson Man: Amos Kendall and the Rise of American Democracy* (Baton Rouge: Louisiana State University Press, 2004), 58.

44. Letters from Amos Kendall to his wife, Jane, January 4, February 14, February 25, 1829, Kendall, *Autobiography*, 277–78 and 285–86.

45. Cole, *Jackson Man*, 123.

46. Amos Kendall to Jane Kendall, April 22, 1829, Kendall, *Autobiography*, 290.

47. Amos Kendall to Jane Kendall, Kendall, *Autobiography*, 297.

48. Cole, *Jackson Man*, 129–30.

49. Amos Kendall to John Branch, November 10, 1829, in *American State Papers*, 3:377.

50. Robert Remini, *Andrew Jackson and the Course of American Freedom, 1822–1832* (New York: Harper and Row, 1981), 185.

51. Mark A. Vargas, "The Military Justice System and the Use of Illegal Punishments as Causes of Desertion in the U.S. Army, 1821–1835," *Journal of Military History* vol. 55, no. 1 (January 1991): 1–19.

52. Richard L. Watson Jr., "Congressional Attitudes toward Military Preparedness, 1829–1835," *Mississippi Valley Historical Review* 34, 4 (March 1948): 613–17.

53. Robert Albion, *Makers of Naval Policy, 1798–1947* (Annapolis, MD: Naval Institute Press, 1980), 631, table 5.

54. Data compiled by the author in, "Born to Rule the Seas."

55. Reynolds's 1839 short story "Mocha Dick" directly influenced Herman Melville's tome *Moby Dick*.

56. Robert Y. Hayne to Secretary of the Navy, January 22, 1829, *Register of Debates in Congress (1828–29)*, appendix, Senate Chamber, 30.

57. Center for Legislative Archives, Senate Committee on Naval Affairs, 20th–26th Congresses, Committee Papers, SEN 20A-D8-D10 (10A-D11), Thompson letter, February 23, 1829, SEN 201A-D9, NARA.

58. *Niles' Weekly Register*, September 1828 to March 1829, vol. 40 (Baltimore: Niles, 1829), 238.

59. For Southard and Hayne's discussion on this topic, see Michael Birkner, *Samuel Southard: Jeffersonian Whig* (Rutherford, NJ: Farleigh Dickinson University Press, 1984), 110–13.

60. As cited in Mary W. M. Hargreaves, *The Presidency of John Quincy Adams* (Lawrence: University Press of Kansas, 1985).

61. *Register of Debates in Congress*, 20th Cong., 2nd Sess., vol. 5 (Washington, DC: Gales and Seaton, 1830), appendix, 35.

62. Center for Legislative Archives, Senate Committee on Naval Affairs, 20th–26th Congresses, Committee Papers, SEN 20A-D8-D10 (10A-D11), Thompson letter, February 23, 1829, SEN 201A-D9, NARA.

63. Congressman Daniel Barnard, February 9, 1831, *Register of Debates in Congress*, 7:149.

64. Senator Robert Hayne, January 4, 1832, *Register of Debates in Congress*, 7:45.

65. Schroeder, "Jacksonian Naval Policy," 126.

Chapter 4

1. For a full account, see Carl P. Borick, *A Gallant Defense: The Siege of Charleston, 1780* (Columbia: University of South Carolina Press, 2003).

2. Hendrik Booraem, *Young Hickory: The Making of Andrew Jackson* (Dallas: Taylor Publishing, 2001), 61.

3. Booraem, *Young Hickory*, 63.

4. See Carl P. Borick, *Relieve Us of This Burthen: American Prisoners of War in the Revolutionary South, 1780–1782* (Columbia: University of South Carolina Press, 2012).

5. Gene A. Smith, *Thomas ap Catesby Jones: Commodore of Manifest Destiny* (Annapolis, MD: Naval Institute Press, 2000), 23–30.

6. Craig Symonds, *Historical Atlas of the U.S. Navy* (Annapolis, MD: Naval Institute Press, 1995), p. 54.

7. Symonds, *Navalists and Antinavalists.*

8. Only oceangoing ships or those that could protect the coastline are included in this study. During the War of 1812, a small fleet was constructed to counter the British invasion from Canada. Despite their moderate successes, they were built only for operations in the Great Lakes and not intended for long-term, postwar use. Jeffersonian gunboats are also not included since they were not commissioned warships.

9. *Laws of the United States in Relation to the Naval Establishment of the Marine Corps* (Washington, DC: Davis and Force, 1826), 72–76.

10. Data compiled in Berube, "Born to Rule the Seas."

11. Letters from ship captains to the secretary of the Navy are a valuable primary source, but the most comprehensive information is available from the "Annual Reports of the Secretary of the Navy" during this period.

12. "Quarterly Return of Receipts of United States Stores and Provisions at Port Mahon

in Charge of Theodore Ludico," Record Group 45, *Reports, Returns, and Estimates Received from Navy Agents*, September 1814–April 1834, NARA.

13. "Annual Report of the Secretary of the Navy, Showing the Condition of the Navy in the Year 1829," reported to Congress December 8, 1829, Naval History and Heritage Command, https://www.history.Navy.mil, accessed May 20, 2018.

14. In addition to Annual Secretary of the Navy Reports, see Record Group 45: Records and Collection of the Office of Naval Records and Library, Record of Boards and Commissions, 1812–1890; Board of Naval Commissioners: Reports of Chief Naval Constructor Samuel Humphreys Concerning the Condition of Naval Vessels, March 7, 1827–April 22, 1834, NARA.

15. Howard Chapelle, *The History of the American Sailing Navy: The Ships and Their Development* (New York: Norton, 1949), 8.

16. Brina J. Agranat, "Thorough and Efficient Repair: Rebuilding in the American Sailing Navy," master's thesis, East Carolina University, 1993, 235.

17. Agranat, "Thorough and Efficient Repair," 238.

18. "Legislative and Executive, of the Congress of the United States. Second session," Document No. 348 "Cost of Fitting a Frigate for Sea, and Annual Expense of Such Vessel," Letter from Secretary of the Navy Samuel Southard to Samuel Smith, Chairman of the Senate Committee of Finance, January 22, 1828, in *American State Papers*, 3:134.

19. Stephen Cassin to John Branch, June 24, 1830, Letters Received from Naval Officers, Letters Received by the Secretary of the Navy from Commissioned Officers below the Rank of Commander, 1802–1884, September 5, 1814–July 5, 1842, 25 vols, microfilm, NARA.

20. Cassin to Branch, Letters Received by the Secretary of the Navy from Commissioned Officers, NARA.

21. Cassin to Branch, Letters Letters Received by the Secretary of the Navy from Commissioned Officers, NARA.

22. Thomas Hamersley to John Branch, March 4, 1830, Navy Courts-Martial, Case 525, Record Group 45, NARA. See also Letters Received by the Secretary of the Navy from Commissioned Officers, September 5, 1814–July 5, 1842, microfilm, NARA.

23. Letter from John Branch to Andrew Jackson, March 4, 1830, Records of General Courts-Martial and Courts of Inquiry of the Navy Department, 1799-1867, Navy Courts-Martial, 1825–1840, Record Group 125, M273, case 525, NARA.

24. Records of General Courts-Martial.

25. Records of General Courts-Martial.

26. Records of General Courts-Martial, case 483.

27. Records of General Courts-Martial, case 484.

28. Records of General Courts-Martial, case 511.

29. Records of General Courts-Martial, case 521.

30. John Wily to Samuel Du Pont, September 29, 1829, Group 9, Du Pont Papers, series B (in file), Box 15 (1814–37).

31. Wiley to Du Pont, September 29, 1829, Du Pont Papers.

32. Records of General Courts-Martial, case 483.

33. Smith, *Thomas ap Catesby Jones*, 34. Other assessments of Creighton continue on 35–37.

34. Court-martial of John Creighton, *Congressional Serial Set* (Washington, DC: Government Printing Office, 1837), 88.

35. Smith, *Thomas Ap Catesby Jones*, 37.

36. Daniel H. Mackay to Samuel F. Du Pont, January 31, 1829, Du Pont Papers.

37. Samuel Du Pont to George Pendergrast, January 23, 1836, Du Pont Papers.

38. Record Group 45, Records Collection of the Office of Naval Records and Library, Records of Boards and Commissions, 1812–90; Board of Naval Commissioners: Estimates and Reports of Expenses of the Navy, 1815–1837, vol. 1, exhibit H, NARA.

39. James Keenz to John Branch, April 7, 1829, Miscellaneous Letters to the Secretary of the Navy, Roll 118, NARA.

40. J. D. Elliott, Navy Yard, Boston, to Commodore [James] Barron, US Navy, Washington, DC, January 14, 1835, James Barron Papers, Swem Library, College of William and Mary, Williamsburg, VA.

41. Anonymous, *A Biographical Notice of Com. Jesse D. Elliott* (Philadelphia: Russell Jarvis, 1835), 326. See also Elliott's letter to the secretary of the Navy, July 3, 1834, Barron Papers, Swem Library, College of William and Mary.

42. *Biographical Notice of Com. Jesse D. Elliott*, 326.

43. *Biographical Notice of Com. Jesse D. Elliott*, 335.

44. This proposal is the first example of what became known as the Base Realignment and Closure Commission (BRAC) in which several post–Cold War commissions sought to save money by closing bases in the United States and globally.

45. John Rodgers to John Branch, March 2, 1836, as reprinted in B. Homans, ed., *Army and Navy Chronicle* vol. 2, no. 22 (June 2, 1836): 337.

46. John Rodgers to John Branch, October 19, 1829, Miscellaneous Letters to the Secretary of the Navy, Roll 118, NARA.

47. Worthy Putnam Sterns, "The Foreign Trade of the United States from 1820 to 1840," *Journal of Political Economy* vol. 8, no. 4 (September 1900): 452–90.

48. "Relative to the Establishment of a Naval Depot," in *American State Papers*, vol. 4.

49. "On the Expediency of Establishing a Navy Yard and Depot at Charleston, South Carolina," June 24, 1834, in *American State Papers*, 4:565–66, 631.

50. Christopher Tomlins, "In Nat Turner's Shadow: Reflections on the Norfolk Dry Dock Affair of 1830–1831," *Labor History* 33 (1992): 498.

51. Andrew Jackson to John Branch, December 24, 1830, Andrew Jackson, *The Papers of Andrew Jackson*, vol. 8, *1830*, ed. by Daniel Feller, Thomas Coens, and Laura-Eve Moss (Knoxville: University of Tennessee Press, 2010), 705.

52. *American State Papers*, 3: Annex E, 7.

53. During Jackson's presidency, South Carolina reacted to the impact of various tariffs by ignoring federal law and threatening secession.

54. The number of ships built under Madison was far greater if platforms in the Great Lakes are considered; however, those were only regional ships for the duration of the war. Their lack of long-term, oceanic capability is therefore the reason they are not included in this study.

55. William T. Allison, Jeffrey Grey, and Janet Valentine, *American Military History: A Survey of Colonial Times to the Present*, 2nd ed. (New York: Routledge, 2016), 82.

56. Lincoln P. Paine, *Warships of the World to 1900* (Boston: Mariner Books, 2000), 56.

57. John S. Wiley to Samuel F. Du Pont, December 9, 1829, Du Pont Papers.

58. Daniel Mackay to Samuel Du Pont, 1 January 1829, Du Pont Papers.

59. Mackay to Du Pont, December 5, 1828, Du Pont Papers.

60. Mackay to Du Pont, December 17, 1828, Du Pont Papers.

61. Mackay to Du Pont, January 1, 1829, Du Pont Papers.

62. Mackay to Du Pont, March 22, 1829, and May 26, 1829, Du Pont Papers.

63. Mackay to Du Pont, August 29, 1829, Du Pont Papers.

64. Extract from a private journal, February 1830, Du Pont Papers.

65. Du Pont Papers, Group 9, Series C, General Files, Box 46, undated, Journals of Trips, 1828–29.

66. Berube and Rodgaard, *Call to the Sea*, 176.

67. Interview with Grant Walker, curator at the US Naval Academy Museum, Annapolis, MD, March 10, 2017. Walker is one of the world's experts on British dockyard models and the Henry Huttleston Rogers Jr. Collection.

68. *Naval and Military Gazette*, London, October 1836; see also *An Apology for English Ship-builders: Showing That It Is Not Necessary the Country Should Look to the Navy for Naval Architects*," 1833, www.jstor.org/stable/60206267.

69. David K. Brown, *Before the Ironclad: Warship Design and Development, 1815–1860* (Annapolis, MD: Naval Institute Press, 2015), 21.

70. William Bainbridge to Samuel Southard, November 18, 1826, Records of Boards and Commissions, Letters Sent to the Secretary of the Navy, April 25, 1815–26, August 1842, vol. 3, NARA.

71. Henry Eckford to John Branch, undated June 1829, Miscellaneous Letters Received by the Secretary of the Navy, 1801–84, Roll 117, VB 255.1.M5, NARA.

72. Henry Eckford to John Branch, June 1829, Miscellaneous Letters Received by the Secretary of the Navy, 1801–84, Roll 117, VB 255.1.M5, NARA.

73. Roy Bolton, *Views of Russia and Russian Works on Paper* (London: Sphinx, 2010), 70.

74. Samuel Humphreys to Isaac Chauncey, February 18, 1838, Letters Received from Naval Constructors and Engineers, microfilm, NARA.

75. Joel Sutherland, Debates in Congress, February 22, 1837, in *Register of Debates in Congress*, vol. 13, parts 1 and 2, 24th Cong., 2nd Sess., 1836–37 (Washington, DC: Gale and Seaton, 1837), 1898–99.

76. Sutherland, Debates in Congress.

77. During the Tripoli War, several commanding officers of sloops were in their mid-twenties such as Charles Stewart, Stephen Decatur, Richard Somers, Isaac Hull, and others. Previously, during the Quasi-War with France, Stewart was still twenty-one when he took command of USS *Syren*.

78. *Register of Debates in Congress*, vol. 13, parts 1 and 2, 24th Cong., 2nd Sess., 1836–37 (Washington, DC: Gale and Seaton, 1837), February 23, 1837. See also Berube, "Budget Battles, Interest Groups and Relevancy," 62–68.

79. *Register of Debates in Congress*, vol. 13, March 2, 1837.

80. Various accounts of the launch can be found in the *Naval Magazine* vol. 2, no. 2 (July 1837): 509–13, and *Daily National Gazette* as recorded in *Niles' Weekly Register*, July 22, 1837, 322.

81. Letter from James Fenimore Cooper to Susan Cooper, September 19, 1837, Cooper, *The Letters and Journals of James Fenimore Cooper*, ed. James Franklin Beard (Cambridge, MA: Belknap Press of Harvard University Press, 1964), 293.

82. *Pennsylvania* was not the last ship of the line built for the Navy, but as stated, it was the last one built with the intention of deploying with squadrons around the world. USS *Ohio* was commissioned in 1838 but was launched in 1820. USS *Vermont* was launched in 1848, completed in 1825, and only served as a receiving ship; it was only commissioned in

1862 to serve during the Civil War. USS *New Hampshire* was completed in 1825 as well as the *Alabama* and commissioned in 1864 as a store ship to relieve the *Vermont*. K. Jack Bauer and Stephen S. Roberts, *Register of Ships of the U.S. Navy, 1775–1900* (New York: Greenwood Press, 1991).

83. Interview, Dr. Robert Remini, House of Representatives, Washington, DC, June 24, 2009.

84. Bernard Brodie, *Seapower in the Machine Age* (Princeton, NJ: Princeton University Press, 1941), 17.

85. Daniel Mackay to Samuel Du Pont, January 1, 1829, Du Pont Papers, Group 9.

86. X, "On Steamers of War," *Naval Magazine*, July 1836, 347.

87. *Military and Naval Magazine of the United States*, March–August 1833, 48.

88. Alexander Hamilton, James Madison, and John Jay, *The Federalist*, with an introduction by Cass R. Sunstein (Cambridge, MA: Harvard University Press, 2009), 64. See also the discussion by John Rohr, "Constitutional Foundations of the United States Navy," *Naval War College Review* 45 (Winter 1992): 68–84.

89. Records of General Courts-Martial, Navy Courts-Martial, 1825–1840, Record Group 125, M273, Cabinet 53A-drawer 5 and Cabinet 54, Rolls 18–35, case 502.

90. Records of General Courts-Martial, case 502.

91. Records of General Courts-Martial, case 502.

92. The 1898 Sampson Board Court of Inquiry concluded the ship was sunk by a mine. The 1911 Vreeland Board Court of Inquiry concluded that an external explosion detonated the magazines. In 1974, Admiral Hyman Rickover's investigation correctly determined that the explosion was caused by combustion from a coal bunker located next to a magazine.

93. John Stevens to Andrew Jackson, March 22, 1830, Jackson, *Papers*, 8:144–45.

94. Stevens to Jackson, March 22, 1830, Jackson, *Papers*, 8:144–45.

95. Alexander Slidell to John Rodgers, March 1835, Record Group 45, Records Collection of the Office of Naval Records and Library, Records of Boards and Commissions, 1812–90; Board of Naval Commissioners: Estimates and Reports of Expenses of the Navy, 1815–1837, NARA.

96. Stevens to Jackson, April 7, 1830, Jackson, *Papers*, 8:179.

97. *American State Papers*, 3: Annex L, 8.

98. John D. Craig to James Barron, November 12, 1833, Barron Collection, College of William and Mary.

99. James Barron to Levi Woodbury, December 30, 1833, Barron Collection, College of William and Mary.

100. Levi Woodbury to James Barron, December 31, 1833, Barron Collection, College of William and Mary.

101. Charles Stewart to James Barron, October 25, 1834, Barron Collection, College of William and Mary.

102. "House of Representatives, Documents, Legislative and Executive of the United States, 23rd Congress, 2nd Session," in *American State Papers*, 3:704.

103. Charles Morris to Mahlon Dickerson, July 29, 1835, Records of Boards and Commissions, Letters Received by Naval Constructors and Engineers, Letters Sent to the Secretary of the Navy, April 25, 1815–August 26, 1842, NARA.

104. James Barron to Andrew Jackson, January 5, 1836, Barron Collection, College of William and Mary.

105. The first *Fulton* was more commonly known as the *Demologos*, the ship that exploded in 1829.

106. Until 1837 his name was Alexander Mackenzie Slidell. He legally inverted his middle and surname to comply with the will of his uncle. For the purposes of this dissertation, he will be referred to by his later legal name except with regard to specific correspondence prior to 1837.

107. *The American in England* (New York: Harper and Brothers, 1835). The 1835 book does not list Slidell as the author, referring to him on the title page as "The Author of 'A Year in Spain.'"

108. Alexander Slidell to John Rodgers, March 1835, Records of Boards and Commissions, 1812–90, Board of Naval Commissioners, Estimates and Reports of Expenses of the Navy, 1815–37, NARA.

109. Brodie, *Seapower in the Machine Age*, 17.

110. A razeed ship was one that had an entire deck removed. For example, a razee frigate was one that was formerly a ship of the line.

111. John Rodgers to Mahlon Dickerson, December 30, 1835, Records of Boards and Commissions, Letters Received from Naval Constructors and Engineers, Letters Sent to the Secretary of the Navy, April 25, 1815–August 26, 1842, NARA.

112. C, untitled letter, *Army and Navy Chronicle*, January 25, 1838, 58.

Chapter 5

1. Sprout and Sprout, *Rise of American Naval Power*, 131.

2. Belohlavek, "Andrew Jackson and the Malaysian Pirates," 19.

3. Andrew Jackson, "Second Annual Message to Congress," December 6, 1830, *Messages of Gen. Andrew Jackson* (Concord: Brown and White, 1837), 93.

4. Although the Webster-Ashburton Treaty falls outside the period covered in this work, readers may wish to consult Kenneth Stevens, *Border Diplomacy: The Caroline and McLeod Affairs in Anglo-American-Canadian Relations, 1837–1842* (Tuscaloosa: University of Alabama Press, 1989).

5. Jackson used the phrase "wide seas" in his 1830 second annual message to Congress and his 1834 sixth annual message to Congress. See Richardson, *Compilation of the Messages and Papers*, 2:526.

6. The Home Squadron, based out of Norfolk, Virginia, was established in 1841 and the Africa Squadron in 1843 to counter illegal slave trading.

7. The effect held true at the beginning of the twenty-first century. The Chinese Navy had not operated globally for five centuries, since the fleet of Zheng He. Modern Somali piracy off the Horn of Africa changed that. As Chinese-flagged ships were threatened, China sent their People's Liberation Army Navy to conduct patrols off the Horn. China has deployed twenty-six flotillas as of August 2017, thus providing it with rich experiences in operating for longer periods in oceans far from their home bases.

8. Thomas Lambert to Lucy Hill Lambert, December 14, 1836, Thomas Lambert Papers, New Hampshire Historical Society, Concord.

9. Jackson, First Annual Message to Congress, December 8, 1829.

10. Dudley Knox, *A History of the United States Navy* (New York: G. P. Putnam's Sons, 1948), 158.

11. Richardson, *Compilation of the Messages and Papers*, 2:443.

12. Remini, *Andrew Jackson and the Course of American Freedom*, 282–83.

13. "Annual Report of the Secretary of the Navy, 1829," in *American State Papers*, December 1, 1829, 3:348.

14. Records of General Courts-Martial, cases 471, 481, and 513 are a few examples.

15. Count Portalis to James Brown, [n.d.], Letters Received by the Secretary of the Navy from Officers below the Rank of Commander, VB255.1, Roll No. 54, November 1–December 31, 1828.

16. Records of General Courts-Martial, case 475.

17. Richardson, *Compilation of the Messages and Papers*, 2:443.

18. Richardson, *Compilation of the Messages and Papers*, 2:443.

19. Richardson, *Compilation of the Messages and Papers*, 2:444.

20. Professor Daniel Feller, editor of the Jackson Papers at the University of Tennessee's Miller Center, briefly discussed this on the center's website, https://millercenter.org/president/jackson/foreign-affairs, accessed September 11, 2017.

21. Andrew Jackson, Sixth Annual Message to Congress, December 1, 1834, in *Messages of Gen. Andrew Jackson*.

22. James Fenimore Cooper, "Comparative Resources of the American Navy," *Naval Magazine*, January 1836, 21.

23. Cooper, "Comparative Resources of the American Navy," 23.

24. James A. Richardson, *A Compilation of the Messages and Papers of the Presidents*, vol. 3, *1833–1841* (Authority of Congress, 1900), 106.

25. Robert Kagan, *Dangerous Nation* (New York: Alfred A. Knopf, 2006), 205.

26. Captain Daniel T. Patterson to the Secretary of the Navy, November 12, 1827, "Account of Destruction of the Turkish Fleet by the English, French, and Russian Fleet at Navarino, and of Piracy by the Greeks in the Mediterranean," in *American State Papers*, Document No. 351, 3:139.

27. Samuel Smith to Samuel Southard, March 17, 1828, Miscellaneous Letters Received by the Secretary of the Navy, 1801–1884, Record Group 45, Roll 117, January 1–March 31, 1829, VB 255.1.M5, NARA.

28. Smith to Southard, March 17, 1828, Miscellaneous Letters Received by the Secretary of the Navy.

29. Daniel Patterson to the Secretary of the Navy, November 12, 1827, Miscellaneous Letters Received by the Secretary of the Navy.

30. Bruce Collins, *War and Empire: The Expansion of Britain, 1790–1830* (New York: Routledge, 2010), 453.

31. Senate Executive Journal, December 28, 1830, 139, http://www.memory.loc.gov/cgi-bin/query/D?hlaw:4:./temp/~ammem_i4Xh, accessed December 8, 2020.

32. Michael Ben Libenson, "American Assistance to the Ottoman Navy in the 1830s and 1840s: A Thesis in History," master's thesis, Pennsylvania State University, 1987, 14.

33. Howard Chapelle, *The History of American Sailing Ships: The Ships and Their Development* (New York: W. W. Norton, 1949), 407.

34. John G. Stephens, *Incidents of Travel in Greece, Turkey, Russia and Poland* (Dublin: Curray, 1839), 191.

35. Alexander W. Hidden, *The Ottoman Dynasty: A History of the Sultans of Turkey* (New York: Hidden, 1912), 308.

36. Richardson, *Compilation of the Messages and Papers*, 3:99.

37. For an account of this conflict, see Charles Napier, *An Account of the War in Portugal between Don Pedro and Don Miguel* (Cambridge: Cambridge University Press, 2013).

38. Richardson, *Compilation of the Messages and Papers*, 2:535–36.

39. Richardson, *Compilation of the Messages and Papers*, 2:445.

40. For more on naval officers conducting diplomacy, the seminal book on the issue is by Long, *Gold Braid and Foreign Relations*.

41. Schroeder, *Matthew Calbraith Perry*, 45.

42. Donald L. Canney, *Africa Squadron: The U.S. Navy and the Slave Trade, 1842–61* (Washington, DC: Potomac Books, 2006), 22–23.

43. Received by the Secretary of the Navy from Commanding Officers of the African Squadron, 1819–1861, Microfilm VB255.1S7A3, Roll 5, July 1832–April 1836. See July 7, 1832, from General P. [Tossen?] to Amos Kendall, and October 22, 1832, from Lieutenant Benjamin Page to Secretary of the Navy Levi Woodbury. Both are at Nimitz Library, US Naval Academy, Annapolis, MD.

44. Douglas R. Egerton, "It's Origin Is Not a Little Curious: A New Look at the American Colonization Society," *Journal of the Early Republic* vol. 5, no. 4 (Winter 1985): 463–80; Andrew Jackson, *The Papers of Andrew Jackson*, vol. 4, *1816–1820*, edited by Harol Moser et al. (Knoxville: University of Tennessee Press, 1994), xxxi.

45. Andrew Jackson to Congress, in *American State Papers*, 3:865.

46. John Berrien to John Branch, *American State Papers*, 3:870. *Executive Documents of the House of Representatives at the Second Session of the Twenty-First Congress*, vol. 3 (Washington: Duff Green, 1831), 9–12.

47. "Annual Report of the Secretary of the Navy, 1830," in *American State Papers*. Vol. 3, *Naval Affairs* (Washington, DC: Gales and Seaton, 1828), 753.

48. Captain John Sloat to Commodore Charles C. Thompson, October 20, 1829, Area File of the Naval Records Collection, 1775–1010, M625, Area 9, Roll 282, April 1814–December 1845, NARA.

49. Finch changed his name to William Compton Bolton in 1833. For continuity purposes, this work will only use his given name.

50. Samuel Southard to John Quincy Adams, January 15, 1829, US Navy Department, Letters Sent by the Secretary of the Navy to the President and Executive Agencies, 1821–1886, Microfilm VB255.1.U72, Roll 1, July 21, 1821–September 30, 1836, NARA.

51. Journals of Thomas A. Dornin, United States Navy, 1826–1855, Microfilm Publication M981, Roll 1, NARA.

52. John C. Jones Jr., US Consul for the Sandwich Islands, to Captain William B. Finch, October 30, 1829, in *American State Papers*, 3:760.

53. Journals of Thomas A. Dornin.

54. Herman Melville, "Journal of Melville's Voyage in a Clipper Ship," *New England Quarterly* vol. 2, no. 1 (January 1929): 120–25.

55. Journals of Thomas A. Dornin, June 19, 1829.

56. Journals of Thomas A. Dornin, July 15, 1829.

57. Journals of Thomas A. Dornin, July 26, 1829.

58. The island is now known as Nei Lingding.

59. Curtis Talmon Henson Jr., "The United States Navy and China, 1839–1861," PhD dissertation, Tulane University, 1965, 1–2.

60. "Report of the Secretary of the Navy," December 6, 1830, in *The National Calendar for 1831*, vol. 9 (Washington, DC: Peter Force, 1831), 321.

61. During the Quasi-War, the US Navy had an insufficient number of ships to conduct operations. Several maritime cities and towns built warships at their own expense to contribute to the war effort. Most were named after the town or county in which they were built or after a notable figure. Salem is in Essex County, Massachusetts.

62. *American State Papers*, 4:154–55.

63. Captain Endicott statement, in *American State Papers*, 4:154.

64. Woodbury to Downes, June 27, 1831, in *American State Papers*, 4:150.

65. Benjamin Armstrong, *Small Boats and Daring Men: Maritime Raiding, Irregular Warfare, and the Early American Republic* (Norman: University of Oklahoma Press, 2019), 151. Armstrong's chapter on the Sumatran expedition should be considered as the authoritative study.

66. Levi Woodbury to John Downes, June 27, 1831, in *American State Papers*, 4:150.

67. Woodbury to Silsbee et al., July 25, 1831, in *American State Papers*, 4:152.

68. Woodbury to Downes, in *American State Papers*, 4:153.

69. Armstrong, *Small Boats and Daring Men*, 159.

70. Belohlavek, "Andrew Jackson and the Malaysian Pirates," 26.

71. A second Sumatran punitive expedition was sent in 1838 in response to attacks but that falls out of the range of this work.

72. Andrew Jackson, Fourth Annual Message to Congress, December 4, 1832, in *Messages of Gen. Andrew Jackson*.

73. Three works briefly mention the *William Penn* incident: Long, *Gold Braid and Foreign Relations*; Nancy Shoemaker, *Native American Whalemen and the World: Indigenous Encounters and the Contingency of Race* (Chapel Hill: University of North Carolina Press, 2015); and Andrew C. A. Jampoler, *Embassy to the Eastern Courts: America's Secret First Pivot toward Asia, 1832–37* (Annapolis, MD: Naval Institute Press, 2015). Jampoler writes that the incident occurred in July, Shoemaker writes that it occurred in September. Long does not provide a month.

74. Long (*Gold Braid and Foreign Relations*, note 55) notes that no villagers were present. Shoemaker (*Native American Whalemen*, note 55), however, writes that the villagers sued for peace.

75. Wilson Heflin, *Herman Melville's Whaling Years* (Nashville, TN: Vanderbilt University Press, 2004), 91–95. Shoemaker, *Native American Whalemen*, 105.

76. Interview, Dr. Robert Remini, House of Representatives, Washington, DC, June 24, 2009.

77. Smith, *Thomas ap Catesby Jones*, 72.

78. Senate Committee on Naval Affairs, February 23, 1829, 20th–26th Congresses, Committee Papers, Center for Legislative Archives, SEN 20A-D8-D10 (10A-D11), NARA.

79. Senate Committee on Naval Affairs, Folder February 23, 1829, 20th–26th Congresses, Committee Papers, Center for Legislative Archives, SEN 20A-D8-D10 (10A-D11), NARA.

80. Senate Committee on Naval Affairs, Folders December 22, 1834, and December 29, 1834, "New York Petition for Expedition to the South Seas," 20th–26th Congresses, Committee Papers, Center for Legislative Archives, SEN 20A-D8-D10 (10A-D11), NARA.

81. Senate Committee on Naval Affairs, Folder December 23, 1834, Citizens of Philadelphia, Exploration Expedition to South Seas, 20th–26th Congresses, Committee Papers, Center for Legislative Archives, SEN 20A-D8-D10 (10A-D11), NARA.

82. Senate Bill 175, March 21, 1835, https://memory.loc.gov/cgi-bin/ampage?collId =llsb&fileName=017/llsb017.db&recNum=359, accessed March 20, 2017.

83. Debate on Exploring Expedition, May 9, 1836, *Congressional Globe*, 337, https:// memory.loc.gov/cgi-bin/ampage?collId=llcg&fileName=112/llcg112.db&recNum=346, accessed March 20, 2017.

84. One of the major sources on the expedition is Wilkes's own autobiography but a better modern account is by Nathaniel Philbrick's *Sea of Glory* (New York: Penguin, 2004).

85. V. F. Boyson, *The Falkland Islands* (Oxford, UK: Clarendon Press, 1924), 90–95.

86. Secretary of the Navy Samuel Southard to Secretary of State Henry Clay, February 19, 1829, US Navy Department, Letters Sent by the Secretary of the Navy to the President and Executive Agencies, 1821–1826, Microfilm VB255.1.U72, Roll 1, July 21, 1821–September 30, 1836, NARA.

87. The United Provinces de la Rio Plata consisted of Buenos Ayres and several adjacent provinces in later Argentina and Uruguay.

88. Craig Evan Klafter, "United States Involvement in the Falkland Islands Crisis of 1831–33," *Journal of the Early Republic* 4, no. 4 (Winter 1984): 409.

89. Gilbert R. Davison, "Public Instrument of Protest," November 23, 1831, Records of the Foreign Service Posts of the Department of State, Buenos Aires, as cited in Klafter, "United States Involvement," 402.

90. Andrew Jackson, Third Annual Message, December 6, 1831, in *Register of Debates in Congress*, vol. 7 (Washington: Gales and Seaton, 1831), 4.

91. Levi Woodbury to John Rodgers, November 29, 1831, Letters Sent by the Secretary of the Navy to Commandants and Navy Agents, as cited in Klafter, "United States Involvement," 411.

92. "Falkland Islands," *Daily National Intelligencer*, May 1, 1832, 2.

93. John Rodgers to Downing, December 23, 1831, Letters Sent by Capt. George W. Rodgers, 1831–32, Letter Books of US Naval Officers, March 1778–July 1908, NARA.

94. Rodgers to Wright, March 23, 1832, Letters Sent by Capt. George W. Rodgers, 1831–32, Letter Books of US Naval Officers, March 1778–July 1908, NARA.

95. "Report of the Secretary of the Navy," in *American State Papers*, 3:214.

96. James Hamilton to John Branch, March 16, 1829, Miscellaneous Letters Received by the Secretary of the Navy, 1801–1884, Roll 117, January 1–March 31, 1829, VB 255.1.M5, NARA.

97. John Rodgers to John Branch, March 16, 1829, Letters Sent to the Secretary of the Navy, April 25, 1815–August 26, 1842, Records of Boards and Commissions, Record Group 45, NARA.

98. Andrew Jackson to John Branch, March 17, 1829, Miscellaneous Letters Received by the Secretary of the Navy, 1801–1884, Roll 117, January 1–March 31, 1829, VB 255.1.M5, NARA.

99. John Rodgers to John Branch, March 26, 1829, Letters Received by the Secretary of the Navy from Officers below the Rank of Commander, VB255.1, Roll No. 54, November 1–December 31, 1828, NARA. See also Records of Boards and Commissions, Letters Sent to the Secretary of the Navy, April 25, 1815–August 26, 1842, NARA.

100. Rodgers to Branch, March 26, 1829, Records of Boards and Commissions, Letters Received by the Secretary of the Navy, NARA.

101. Jackson, *Papers*, 8:45.

102. Daniel H. Mackay to Samuel Du Pont, December 5, 1828, Group 9, Du Pont Papers, Series B (in file), Box 15 (1814–37).

103. Mackay to Du Pont, December 5, 1828, Du Pont Papers.

104. Daniel H. Mackay to Samuel Du Pont, January 31, 1829, Du Pont Papers.

105. Proceedings of the House of Representatives, February 18, 1830, http://memory.loc.gov/cgi-bin/query/D?hlaw:19:./temp/~ammem_Y8V6, accessed December 8, 2020.

106. Proceedings of the House of Representatives.

107. Proceedings of the House of Representatives.

108. See Richard A. Katula, *The Eloquence of Edward Everett, America's Finest Orator* (New York: Peter Lang, 2010).

109. Katula, *Eloquence of Edward Everett.*

110. Elmer Flaccus, "Commodore David Porter and the Mexican Navy," *Hispanic American Historical Review* 34, no. 3 (August 1954): 366.

111. Flaccus, "Commodore David Porter," 367.

112. Flaccus, "Commodore David Porter," 372.

113. J. Fred Rippy, *Joel R. Poinsett: Versatile American* (Durham NC: Duke University Press, 1935), 129.

114. Jackson, *Papers*, 7:498–99.

115. For a complete discussion on the Texas and Mexico issue, see David M. Pletcher, *The Diplomacy of Annexation: Texas, Oregon, and the Mexican War* (Columbia: University of Missouri Press, 1973).

116. Bruce Vandervort, *Indian Wars of Mexico, Canada, and the United States, 1812–1900* (New York: Routledge, 2006), 130.

117. Robert Remini, *Andrew Jackson and His Indian Wars* (New York: Viking, 2001), 33.

118. Mahlon Dickerson to Alexander Dallas, October 29, 1835, Press Copies of Reports Sent Concerning Navy and Marine Corps Service in the Second Seminole War, 1835–42, Part I: Instructions to Officers of the United States Navy Cooperating with the Land Forces of the US during the Florida War, 1835–42, Record Group 45, NARA.

119. Dickerson to Dallas, January 30, 1836, Press Copies of Reports.

120. Dickerson to Dallas, January 9, 1836, Press Copies of Reports.

121. Dickerson to Dallas, January 21, 1836, Press Copies of Reports.

122. Official correspondence reported USS *Concord*'s arrival on April 8, 1836. However, Captain Mix's diary stated he arrived off Tampa Bay on March 30. Private journal of Captain Mix kept on board of the *Concord* sloop of war, Special Collections, University of South Florida Tampa Library.

123. Dallas to Dickerson, January 15, 1836, Press Copies of Reports.

124. James Mooney, ed., *Dictionary of American Naval Fighting Ships*, vol. 6 (Washington, DC: Government Printing Office, 1976), 402.

125. Jackson's order is incorporated in Dickerson's letter to Dallas. Dickerson to Dallas, May 25, 1836, Press Copies of Reports.

126. Dickerson to Dallas, May 25, 1836, Press Copies of Reports.

127. Dallas to Dickerson, October 8, 1836, and January 19, 1837, Press Copies of Reports.

128. George E. Buker, "Lieutenant Levin M. Powell, U.S.N., Pioneer of Riverine Warfare," *Florida Historical Quarterly* 47 (January 1969): 267.

129. Knox, *History of the United States Navy*, 158.

130. Archibald Henderson to Mahlon Dickerson, May 11, 1837, Record Group 80,

General Records of the Navy, Letters to the Commandant and Other Officers of the Marine Corps, April 1804–November 1886, vol. 16, NARA.

131. Henderson to Mahlon Dickerson, May 11, 1837, Record Group 80.

132. Andrew Jackson, Eighth Annual Message to Congress, December 5, 1836, in *Messages of Gen. Andrew Jackson.*

133. *Register of Debates in Congress*, vol. 14, pt. 1, 25th Cong., 1st Sess., 1837 (Washington, DC: Gales and Seaton, 1860), 640.

134. *Register of Debates in Congress*, 14:642.

135. Schroeder, "Jacksonian Naval Policy," 123.

136. Klafter, "United States Involvement," 408.

137. The most extensive firsthand account of the squadron's first three years is William S. W. Ruschenberger, *A Voyage Round the World Including an Embassy to Muscat and Siam in 1835, 1836, and 1837* (Philadelphia: Carey, Lea and Blanchard, 1838), Nimitz Library, US Naval Academy, Special Collections G 440.R96. Some other records may be found at the Area File of the Naval Records Collection, 1775–1910, M625, Area 10, Roll 345, 1798–1851, NARA.

Chapter 6

1. Charles Stewart to Secretary of the Navy John Branch, March 8, 1830, appendix to Secretary of the Navy John Branch to the US Senate, March 23, 1830, "On the Expediency of Dispensing with the Marine Corps as Part of the Armed Equipment of a Vessel-of-War," in *American State Papers*, 3:560–69.

2. James E. Valle, *Rocks and Shoals: Naval Discipline in the Age of Fighting Sail* (Annapolis: Naval Institute Press, 1996), 3.

3. Records of General Courts-Martial, case 640.

4. Records of General Courts-Martial, case 651.

5. Valle, *Rocks and Shoals*, 91.

6. Berube and Rodgaard, *Call to the Sea*, 173.

7. Harold Langley, *Social Reform in the United States Navy, 1708–1862* (Urbana: University of Illinois Press, 1967), 174.

8. Records of General Courts-Martial, case 660.

9. Valle, *Rocks and Shoals*, 91.

10. Valle, *Rocks and Shoals*, 57.

11. A raze was a ship that had been reduced by one deck. For example, a ship of the line might have had an entire deck removed. The loss in the number of guns it carried would likely make it a frigate.

12. "Jackson and Lieut. Randolph," *New York Times*, April 17, 1893.

13. John M. Belohlavek, "Assault on the President: The Jackson-Randolph Affair of 1833," *Presidential Studies Quarterly* vo. 12, no. 3 (Summer 1982): 362.

14. "The Assault of Lieut. Randolph," *Virginia Free Press*, May 16, 1833, affirms like other newspaper accounts that Randolph had been exonerated by the court of inquiry. A reading of the actual inquiry is more nuanced. Randolph was criticized for failing to manage the accounts such as not taking receipts. He was not intentionally defrauding the government.

15. Records of General Courts-Martial, case 579.

16. Records of General Courts-Martial, case 579.

17. Records of General Courts-Martial, case 579.

18. Records of General Courts-Martial, case 579.

19. Most accounts state that the attack occurred while the *Sydney* was in Alexandria about to depart for Fredericksburg; Belohlavek writes that it occurred on May 6 on the president's return from Fredericksburg, "Assault on the President," 361.

20. Belohlavek, "Assault on the President," 362.

21. Washington Irving to Peter Irving, May 17, 1833, Irving, *Complete Works*, 762.

22. Patrick J. Novotny, *The Press in American Politics, 1787–2012* (Santa Barbara: Praeger, 2014), 33.

23. *Globe* (Washington, DC), May 8, 1833.

24. *Globe*, May 8, 1833.

25. "Death of an Old Citizen," *Daily National Intelligencer*, April 27, 1869.

26. House Document, no. 116, 21st Cong., December 4, 1838, in *American State Papers*, vol. 3. The pardon is also discussed in Robert Remini, *Andrew Jackson and the Course of American Democracy, 1833–1845* (Baltimore: Johns Hopkins University Press, 1984), 62.

27. Chisholm, *Waiting for Dead Men's Shoes*, 134.

28. Leland Pearson Lovette, *School of the Sea: The Annapolis Tradition in American Life* (New York: Frederick A. Stokes, 1941), 258.

29. William A. Duer to Isaac Chauncey, February 3, 1830, Center for Legislative Archives, HR21A-D17.5, NARA.

30. Isaac Chauncey to John Branch, February 4, 1830, Center for Legislative Archives, HR21A-D17.5, NARA.

31. Email exchange between the author and Jocelyn Wilk, Columbia University Special Collections, June 10, 2011.

32. "Annual Report of the Secretary of the Navy, December 1830," 3:758.

33. Register of Debates, Senate, 24th Cong., 1st Sess., 1453, http://memory.loc.gov/cgi-bin/ampage?collId=llrd&fileName=023/llrd023.db&recNum=90, accessed December 8, 2020.

34. *American State Papers*, 4:884.

35. *American State Papers*, 4:884.

36. Callahan, *List of Officers*.

37. The authoritative author on this subject is Harold Langley, *The History of Medicine in the Early U.S. Navy* (Baltimore: Johns Hopkins Press, 2000).

38. Data compiled by the author from annual reports of the secretaries of the Navy in the course of research for "Born to Rule the Seas."

39. Langley, *History of Medicine*, 23.

40. Andre Sobocinski, "A Navy Surgeon on Olympus," *Navy Medicine* vol. 96, no. 5 (2005): 28–31 (Bethesda, MD: Navy Medical Department).

41. There is no known relation between William and Clara Barton.

42. Capt. Edmund P. Kennedy to Commodore John Rogers, Pres. Navy Board, Washington, DC, January 14, 1824; James Barron to Captain Lewis Warrington, March 25, 1824; Edmund P. Kennedy to Comdr. James Barron, August 8, 1827, James Barron Papers, Mss. 65 B27, College of William and Mary.

43. Ellis Hudson, "Quinine and Atabrine: Development and Present Application," *United States Naval Medical Bulletin* vol. 45, no. 1 (1941): 57 (Washington, DC: General Printing Office).

44. James Bradford, *Command under Sail: Makers of the American Naval Tradition, 1775–1850* (Annapolis, MD: Naval Institute Press, 1985), 234.

45. Garrett J. Pendergrast to Samuel F. Du Pont, January 5, 1824, Louisville, KY, Du Pont Papers, Group 9. Pendergrast also commanded *Cumberland* at the outset of the Civil War.

46. Records of General Courts-Martial, case 426.

47. Records of General Courts-Martial, case 426.

48. Records of General Courts-Martial, case 426.

49. Records of General Courts-Martial, case 490.

50. Records of General Courts-Martial, case 490.

51. Records of General Courts-Martial, case 490.

52. Records of General Courts-Martial, case 490.

53. "Annual Report of the Secretary of the Navy, December 1830."

54. Records of General Courts-Martial, case 630.

55. The Liverpool School of Tropical Medicine was founded in 1898.

56. Samuel Southard to John Quincy Adams, February 9, 1829, US Navy Department, Letters Sent by the Secretary of the Navy to the President and Executive Agencies, 1821–1886, Microfilm VB255.1.U72, Roll 1, July 21, 1821–September 30, 1836, NARA.

57. Robert Hayne, February 16, 1830, Center for Legislative Archives, SEN 21A-D11, February 16, 1830, NARA.

58. *Globe* (Indianapolis), Saturday, January 14, 1832.

59. Howe, *What Hath God Wrought*, 180.

60. Howe, *What Hath God Wrought*, 186.

61. Howe, *What Hath God Wrought*, 194.

62. Benjamin Tappan also later served as a US senator from 1839 to 1845.

63. Tappan, *Life of Arthur Tappan*, 91.

64. Ronald N. Satz, *American Indian Policy in the Jackson Era* (Norman: University of Oklahoma Press, 1975), 21.

65. Clifford M. Drury, *The History of the Chaplain Corps, United States Navy*, vol. 1, *1778–1939* (Washington, DC: Bureau of Naval Personnel, 1948), 51.

66. Drury, *History of the Chaplain Corps*, 52.

67. Drury, *History of the Chaplain Corps*.

68. Drury, *History of the Chaplain Corps*.

69. Enoch Cobb Wines, *Two and a Half Years in the Navy: A Journal of a Cruise in the Mediterranean and Levant on Board the U. S. Frigate* Constellation *in the Years 1829, 1830, and 1831* (Philadelphia: Carey and Lea, 1832), 48.

70. Langley, *Social Reform*, 61.

71. *Sailors Magazine and Naval Journal* vol. 1 (1828): ii.

72. Reverend Charles S. Stewart, *Sailors Magazine and Naval Journal* vol. 1 (1829): iii.

73. American Spectator, "The Navy," *Sailors Magazine and Naval Journal* vol. 3 (1830): 90.

74. American Spectator, "Navy."

75. Berube and Rodgaard, *Call to the Sea*, 203–4.

76. Drury, *History of the Chaplain Corps*, 1.

77. Charles Rockwell, *Sketches of Foreign Travel and Life at Sea: Including a Cruise Aboard a Man of War*, vol. 2 (New York: Tappan and Dennet, 1842), 392.

78. Research on US Navy courts-martial compiled by author at National Archives, Washington, DC.

79. "Annual Report of the Secretary of the Navy," in *American State Papers*, 3:8.

80. Drury, *History of the Chaplain Corps*, 47.

81. Drury, *History of the Chaplain Corps*, 117.

82. Drury, *History of the Chaplain Corps*, 43.

83. Data compiled from secretary of the Navy annual reports, which listed personnel by rank and duty.

84. Drury, *History of the Chaplain Corps*, 81.

85. H. Lawrence Martin and Timothy J. Demy, *Sea Padres* (Washington, DC: Office of the Chief of Chaplains, 1984), 7.

86. Both Colton and Ryland were active members of the United States Colonization Society, which sought to repatriate freed slaves to Africa.

87. "Application of the Chaplains of the Navy for an Increase of Pay," January 18, 1832, in *American State Papers*, 4:84–85.

88. Drury, *History of the Chaplain Corps*, 47.

89. Slaves were not technically "free" labor since there was a cost initially of acquiring them and then, if viewed in terms of a commodity as they were in the South, there were the operational and maintenance costs—housing, food, and so on.

90. Spencer Tucker, ed., *Enduring Controversies in Military History* (Santa Barbara, CA: ABC-CLIO, 2017), 365.

91. Unknown author, "Colored Sailors Home," *Sailor's Magazine*, 1849, 253.

92. Steedman, *Memoir and Correspondence*, 27.

93. Steedman, *Memoir and Correspondence*, 40–41.

94. Steedman, *Memoir and Correspondence.*

95. John Rodgers to Levi Woodbury, February 8, 1833, Records of Boards and Commissions, Letters Sent to the Secretary of the Navy, April 25, 1815–August 26, 1842, NARA.

96. Norfolk was not the only center for this enslavement work. The Army also used enslaved people. See Mark A. Smith: "Engineering Slavery: The U.S. Army Corps of Engineers and Slavery at Key West," *Florida Historical Quarterly* vol. 86, no. 4 (Spring 2008): 505–10.

97. Tomlins, "In Nat Turner's Shadow," 500.

98. John Rodgers to Levi Woodbury, October 15, 1831, Letters Sent to the Secretary of the Navy, April 25, 1815–August 26, 1842, Records of Boards and Commissions, Letters Sent to the Secretary of the Navy, April 25, 1815–August 26, 1842.

99. John Rodgers to Levi Woodbury, February 8, 1833, Records of Boards and Commissions, Letters Sent to the Secretary of the Navy, April 25, 1815–August 26, 1842, vol. 4.

100. Acting Secretary John Boyle to Archibald Henderson, September 17, 1831, Record Group 80, General Records of the Department of the Navy, Letters to the Commandant and Other Officers of the Marine Corps, April 1804–November 1886, NARA.

101. Tomlins, "In Nat Turner's Shadow." For further reference, see Loammi Baldwin Papers, (Boxes 8, 17–23), Harvard Business School, Baker Library, Boston.

102. For more on the Turner rebellions, see David Allmendinger, *Nat Turner and the Rising in Southampton County* (Baltimore: Johns Hopkins University Press, 2014); Judith Edwards, *Nat Turner's Slave Rebellion in American History* (New York: Enslow, 2000); Russell Nye, "Freedom Road: Nat Turner," *Centennial Review of Arts and Science* vol. 1, no. 3 (Summer 1957): 242–63.

103. Barbara Brooks Tomblin, *Bluejackets and Contrabands: African Americans and the Union Navy* (Lexington: University of Kentucky Press, 2009), 17.

104. Mark A. Smith, *Engineering Security: The Corps of Engineers and Third System Defense Policy, 1815–1861* (Tuscaloosa: University of Alabama Press, 2009), 169–71.

105. Rodgers to Woodbury, October, 15, 1831, Board of Navy Commissioners, Circular to Commandants of Naval Shipyards, March 17, 1817: "Abuses having existed in some of the Navy yards by the introduction of improper Characters for improper purposes, The Board of Navy Commissioners have deemed it necessary to direct that no Slaves or Negroes, except under extraordinary Circumstances, shall be employed in any Navy yard in the United States, & in no case without the authority from the Board of Navy Commissioners."

106. For accounts of the Washington Navy Yard strike of 1835, see Jefferson Morley *Snow-Storm in August: The Struggle for American Freedom and Washington's Race Riot of 1835* (New York: Anchor, 2012); David Williams, *A People's History of the Civil War* (New York: New Press, 2005); as well as contemporary newspaper accounts.

107. See W. J. Rorabaugh, *The Alcoholic Republic: An American Tradition* (New York: Oxford University Press, 1979).

108. Howe, *What Hath God Wrought*, 167.

109. Major General Alexander Macomb to Secretary of War John Eaton, House of Representatives, 21st Cong., 1st Sess., January 11, 1830, https://collections.wip.nlm.nih.gov/ocr/nlm:nlmuid-101609859-bk.

110. Richard Rush also served as the vice presidential nominee under John Quincy Adams during the 1828 election in which they were defeated by Andrew Jackson. Several years later, Jackson would appoint Rush as a commissioner to England for funds creating the Smithsonian Institution. He would later be appointed minister to France by James K. Polk, Jackson's protégé.

111. Howe, *What Hath God Wrought*, 167.

112. House of Representatives, 21st Cong., 1st Sess., Document No. 22, War Department Letter on "Ardent Spirits—Army and Navy," https://collections.wip.nlm.nih.gov/ocr/nlm:nlmuid-101609859-bk.

113. House of Representatives, War Department Letter on "Ardent Spirits—Army and Navy."

114. House of Representatives, War Department Letter on "Ardent Spirits—Army and Navy."

115. House of Representatives, War Department Letter on "Ardent Spirits—Army and Navy."

116. George Gibson to John Eaton, House of Representatives, 21st Congr., 1st Sess., January 6, 1830, https://collections.wip.nlm.nih.gov/ocr/nlm:nlmuid-101609859-bk.

117. Although Heerman spent most of his career in New Orleans, he was no stranger to direct naval action having volunteered to join Lieutenant Stephen Decatur's mission aboard the ketch *Intrepid* to recapture and destroy the USS *Philadelphia* in Tripoli Harbor in 1804.

118. Lewis Heerman to John Eaton, House of Representatives, 21st Cong., 1st Sess., June 25, 1829, https://books.google.com/books?id=G3QZAAAAIAAJ.

119. Heerman to Eaton, House of Representatives.

120. Heerman to Eaton, House of Representatives.

121. William Barton to John Eaton, House of Representatives, 21st Cong., 1st Sess., September 14, 1829, https://books.google.com/books?id=G3QZAAAAIAAJ.

122. As a board member, Barton reviewed the first applications of surgeon's mates in 1824 and became commanding officer of the Naval Hospital in Norfolk in 1830. A decade later, he was appointed as the first head of the Bureau of Medicine and Surgery.

123. One of the main characters in Shakespeare's *Othello*.

124. Thomas Harris to John Eaton, House of Representatives, 21st Cong., 1st Sess., September 12, 1829, https://books.google.com/books?id=G3QZAAAAIAAJ.

125. Harris would later succeed Barton as head of the Bureau of Naval Medicine and Surgery.

126. Harris to Eaton, House of Representatives.

127. Harris to Eaton, House of Representatives.

128. Schroeder, *Matthew Calbraith Perry*), 10.

129. Record Group 125, M273, Roll 22, vol. 20, case 478, NARA.

130. Jackson, *Papers*, 8:164–65.

131. Record Group 125, M273, Roll 22, Vol. 20, case 481, NARA.

132. Record Group 125, M273, Roll 22, Vol. 20, case 487, NARA.

133. Record Group 125, M273, Roll 22, Vol. 20, case 497, NARA.

134. Record Group 125, M273, Roll 22, Vol. 20, case 500, NARA.

135. Record Group 125, M273, Roll 22, Vol. 20, case 472, NARA.

136. Record Group 125, M273, Roll 22, Vol. 20, cases 476 and 477, NARA.

137. Record Group 125, M273, Roll 22, Vol. 20, case 496, NARA.

138. Levi Woodbury to Campbell P. White, 23rd Cong., 1st Sess., Document no. 486, June 10, 1834, in *American State Papers*, vol. 4.

139. "On the Expediency of Dispensing with the Marine Corps," 3:560–69.

140. "On the Expediency of Dispensing with the Marine Corps," 3:560–69.

141. Joseph G. Dawson, "With Fidelity and Effectiveness: Archibald Henderson's Lasting Legacy to the U.S. Marine Corps," *Journal of Military History* 62.4 (October 1998): 731.

142. Berube and Rodgaard, *Call to the Sea*, 208.

143. "On the Expediency of Dispensing with the Marine Corps," 3:560–69.

144. Captain Charles Broome to Levi Woodbury, January 28, 1834, Record Group 80, General Records of the Department of the Navy, Letters to the Commandant and Other Officers of the Marine Corps, April 1804–November 1886, NARA.

145. Mahlon Dickerson to Archibald Henderson, March 11, 1835, Record Group 80, General Records of the Department of the Navy, Letters to the Commandant.

146. Joseph G. Dawson, "With Fidelity and Effectiveness," 735.

147. Mahlon Dickerson to Archibald Henderson, January 29, 1835, Record Group 80, General Records of the Department of the Navy, Letters to the Commandant.

148. Berube, "Born to Rule the Seas," 261.

149. Archibald Henderson to Mahlon Dickerson, December 8, 1835, Record Group 80, General Records of the Department of the Navy, Letters to the Commandant.

150. Archibald Henderson to Mahlon Dickerson, December 24, 1835, Record Group 80, General Records of the Department of the Navy, Letters to the Commandant.

Conclusion

1. Records of the Naval Lyceum including account ledgers, artifact donations, and visitor logs are in the possession of the Naval Academy Museum in Annapolis, MD.

2. Julius W. Pratt, "The Origin of Manifest Destiny," *American Historical Review*, 4 (1927): 795.

Bibliography

Primary Sources

Manuscript Sources

College of William and Mary, Special Collections Research Centre, Swem Library, Williamsburg, VA

James Barron Collection

Columbia University Archives, New York

Hagley Museum and Library, Brandywine, DE

Samuel Du Pont Papers

Harvard Business School, Baker Library, Boston

Loammi Baldwin Papers

Library of Congress, Washington, DC

Massachusetts Historical Society, Boston

Adams-Waterhouse Letters
Everett-Peabody Papers
Washington Irving Papers
Enoch Preble Diary and Memorandum Book
George H. Preble Papers
US Naval Hospital Letterbook
Charles Henry Wheelwright Letters

National Archives and Records Administration, Washington, DC

Board of Naval Commissioners: Estimates and Reports of Expenses of the Navy, 1815–1837

Board of Naval Commissioners: Reports of Chief Naval Constructor Samuel Humphreys Concerning the Condition of Naval Vessels, March 7, 1827–April 22, 1834

Center for Legislative Archives

House Committee on Naval Affairs
Senate Committee on Naval Affairs

Journal of a Cruise in the United States Frigate *Brandywine* and *Boxer*, April 28, 1834–February 13, 1837, Commander David Deacon.

Journals Kept on Board the US Schooner *Experiment* and the US Schooner *Nautilus*, August 4, 1835–November 8, 1841, Lieutenant Commander George Blake

Journals of John Julius Guthrie, July 18, 1834–June 23, 1858
Journal of Lieutenant Thomas A. Dornin

Journal of Midshipman Abram D. Harrel on North Carolina Journal of Midshipman William P. Gamble

Letter Books of U.S. Naval Officers, March 1778–July 1908

Laws of the United States in Relation to the Naval Establishment of the Marine Corps, Washington, DC: Davis and Force, 1826

Letters from the Commandant of the Marine Corps, January 1828–December 1886

Letters Received by the Secretary of the Navy from Captains, 1802–84

Letters Received by the Secretary of the Navy from Commanders, 1802–84

Letters Received by the Secretary of the Navy from Commissioned Officers below the Rank of Commander, 1802–84

Letters Received from Naval Constructors and Engineers

Letters Received from Naval Officers, September 5, 1814–July 5, 1842

Letters to the Commandant and Other Officers of the Marine Corps, April 1804–November 1886

Letters Sent by the Secretary of the Navy to Officers, 1798–1871

Letters Sent by the Secretary of the Navy to the President and Executive Agencies, 1821–86

Logs and Journals Kept by Lt. William Mervine, Commanding the USS *Experiment, Natchez, Cyane, Savannah, Mississippi, Colorado,* and *Rhode Island*, April 22, 1832–September 22, 1861

Miscellaneous Letters Received by the Secretary of the Navy, 1801–84

Miscellaneous Letters to the Secretary of the Navy, 1801–84

Press Copies of Reports Sent Concerning Navy and Marine Corps Service in the Second Seminole War, 1835–42

Received by the Secretary of the Navy from Commanding Officers of the African Squadron, 1819–1861

Record Group 45

Area File of the Naval Records Collection, 1775–1910

Correspondence of Rear Adm. William Mervine, July 1836–August 1868

Courts-Martial and Courts of Inquiry of the Navy Department, 1799–1867

Record Group 80

Record Group 125

Records Collection of the Office of Naval Records and Library Records of Boards and Commissions, 1812–90, Journal of Meetings, April 25, 1815–September 3, 1842

Records of Boards and Commissions, 1812–90, Letters Received from Naval Constructors and Engineers, May 9, 1815–July 5, 1842

Records of Boards and Commissions, Letters Sent to the Secretary of the Navy, April 25, 1815–August 26, 1842

Reports of Officers of the United States Navy Whilst Co-operating with the Land Forces of the United States during the Florida War

Reports, Returns, and Estimates Received from Navy Agents, September 1814–April 1834

New Hampshire Historical Society, Concord
Thomas Lambert Papers

University of South Florida, Tampa Library Special Collections
Private Journal of Captain Mervine Mix

US Naval Academy, Nimitz Library, Special Collections and Archives, Annapolis, MD
John Lenthall Papers

US Naval Academy Museum, Annapolis, MD
Records of the Naval Lyceum

Original Sources

American State Papers. Vol. 3, *Naval Affairs*. Washington, DC: Gales and Seaton, 1861.

American State Papers. Vol. 4, *Naval Affairs*. Washington, DC: Gales and Seaton, 1861.

"Annual Report of the Secretary of the Navy." In *American State Papers*. Vol. 3, *Naval Affairs*. Washington, DC: Gales and Seaton, 1828.

"Annual Report of the Secretary of the Navy, Showing the Condition of the Navy in the Year 1829." Reported to Congress December 8, 1829. Naval History and Heritage Command. https://www.history.navy.mil.

"Annual Report of the Secretary of the Navy, 1830." In *American State Papers*. Vol. 3, *Naval Affairs*. Washington, DC: Gales and Seaton, 1828.

"Annual Report of the Secretary of the Navy, 1831." In *American State Papers*. Vol. 3, *Naval Affairs*. Washington, DC: Gales and Seaton, 1828.

"Annual Report of the Secretary of the Navy, 1832." In *American State Papers*. Vol. 3, *Naval Affairs*. Washington, DC: Gales and Seaton, 1828.

Congressional Globe. Washington, DC: John C. Rives.

Congressional Record. Washington, DC: US Government Printing Office.

Congressional Serial Set. Washington, DC: Government Printing Office, 1837.

Executive Documents of the House of Representatives at the Second Session of the Twenty-First Congress. Vol. 3. Washington, DC: Duff Green, 1831.

Jackson, Andrew. Eighth Annual Message to Congress, December 5, 1836. In *Messages of Gen. Andrew Jackson*. Concord, NH: Brown and White, 1837.

———. Farewell Address, March 4, 1837. In *Messages of Gen. Andrew Jackson*. Concord, NH: Brown and White, 1837.

———. Fifth Annual Message to Congress, December 1833. In *Messages of Gen. Andrew Jackson*. Concord, NH: Brown and White, 1837.

———. First Annual Message to Congress, December 8, 1829. In *Messages of Gen. Andrew Jackson*. Concord, NH: Brown and White, 1837.

———. First Inaugural Address, March 4, 1829. http://avalon.law.yale.edu/19th_century/jackson1.asp.

———. Fourth Annual Message to Congress, December 4, 1832. In *Messages of Gen. Andrew Jackson*. Concord, NH: Brown and White, 1837.

———. Second Annual Message to Congress, December 6, 1830. In *Messages of Gen. Andrew Jackson*. Concord, NH: Brown and White, 1837.

———. Second Inaugural Address, March 4, 1833. https://avalon.law.yale.edu/19th_century/jackson2.asp.

———. Seventh Annual Message to Congress, December 7, 1835. In *Messages of Gen. Andrew Jackson*. Concord, NH: Brown and White, 1837.

———. Sixth Annual Message to Congress, December 1, 1834. In *Messages of Gen. Andrew Jackson*. Concord, NH: Brown and White, 1837.

Register of Debates in Congress. Vol. 5. Washington, DC: Gales and Seaton, 1830.

Register of Debates in Congress. Vol 7. Washington, DC: Gales and Seaton, 1831.

Register of Debates in Congress. Vol. 13, parts 1 and 2. 24th Congress, 2nd Session, 1836–37. Washington, DC: Gale and Seaton, 1837.

Register of Debates in Congress. Vol. 14, part. 1. 25th Congress, 1st Session, 1837. Washington, DC: Gales and Seaton, 1860.

"Report of the Secretary of the Navy." December 6, 1830. In *The National Calendar for 1831*. Vol 9. Washington, DC: Peter Force, 1831.

United States, Department of the Navy. *Report of the Secretary of the Navy, 1889*. Washington, DC: Government Printing Office,1890.

US Department of Commerce. *Historical Statistics of the United States*. Washington, DC: Government Printing Office, 1949.

Books

Adams, John Quincy. *Memoirs of John Quincy Adams: Comprising Portions of His Diary from 1795–1848*. Vol. 5. Edited by Charles Francis Adams. Philadelphia: Lippincott, 1875.

Ambrose, Stephen. *Duty, Honor, Country. A History of West Point*. Baltimore: Johns Hopkins University Press, 1966.

Callahan, Edward, ed. *List of Officers of the Navy of the United States and of the Marine Corps from 1775 to 1900*. New York: L. R. Hamersly, 1901.

Clay, Henry. *The Papers of Henry Clay*. Vol. 8, *Candidate, Compromiser, Whig, March 5, 1829–December 31, 1836*. Edited by Robert Seager II and Melba Porter Hay. Lexington: University of Kentucky Press, 1984.

Cooper, James Fenimore. *The Letters and Journals of James Fenimore Cooper*. Edited by James Franklin Beard. Cambridge, MA: Belknap Press of Harvard University Press, 1964.

Dallas, Francis Gregory. *The Papers of Francis Gregory Dallas, United States Navy, Correspondence and Journal, 1837–1859*. Edited by Gardner W. Allen. New York: Naval Historical Society, 1917.

Jackson, Andrew. *The Papers of Andrew Jackson*. Vol. 3, *1814–1815*. Edited by Harold D. Moser, David R. Hoth, Sharon Macpherson, and John H. Reinbold. Knoxville: University of Tennessee Press, 1991.

———. *The Papers of Andrew Jackson*. Vol. 4, *1816–1820*. Edited by Harold D. Moser, David R. Hoth, and George H. Hoemann. Knoxville: University of Tennessee Press, 1994.

———. *The Papers of Andrew Jackson*. Vol. 7, *1829*. Edited by Daniel Feller, Harold D. Moser, Laura-Eve Moss, and Thomas Coens. Knoxville: University of Tennessee Press, 2007.

———. *The Papers of Andrew Jackson*. Vol. 8, *1830*. Edited by Daniel Feller, Thomas Coens, and Laura-Eve Moss. Knoxville, University of Tennessee Press, 2010.

———. *The Papers of Andrew Jackson*. Vol. 9, *1831*. Edited by Daniel Feller, Laura-Eve

Moss, Thomas Coens, and Erik B. Alexander. Knoxville: University of Tennessee Press, 2013.

Kendall, Amos. *Autobiography of Amos Kendall.* Edited by William Stickney. Boston: Lee and Shepherd, 1872.

Paulding, James Kirke. *The Letters of James Kirke Paulding.* Edited by Ralph M. Aderman. Madison: University of Wisconsin Press, 1962.

Rockwell, Charles. *Sketches of Foreign Travel and Life at Sea: Including a Cruise Aboard a Man of War.* Vol. 2. New York: Tappan and Dennet, 1842.

Ruschenberger, William S. W. *A Voyage Round the World Including an Embassy to Muscat and Siam in 1835, 1836, and 1837.* Philadelphia: Carey, Lea and Blanchard, 1838.

Sands, Benjamin F. *From Reefer to Rear-Admiral: Reminiscences and Journal Jottings of Nearly Half a Century of Naval Life.* New York: Frederick A. Stokes, 1899.

Steedman, Charles. *Memoir and Correspondence of Charles Steedman, Rear Admiral, United States Navy, with His Autobiography and Private Journals, 1811–1890.* Cambridge, MA: Riverside Press, 1912.

Tappan, Lewis. *The Life of Arthur Tappan.* New York: Hurd and Houghton, 1870.

Trollope, Frances. *Domestic Manners of the Americans.* London: Whittaker, Teacher, 1832.

Warriner, Francis. *Cruise of the U.S. Frigate Potomac Round the World, 1831–34.* New York: Leavitt, Lord, 1835.

Wines, Enoch Cobb. *Two and a Half Years in the Navy: A Journal of a Cruise in the Mediterranean and Levant on Board the U. S. Frigate* Constellation *in the Years 1829, 1830, and 1831.* Philadelphia: Carey and Lea, 1832.

Periodical Magazines and Newspapers

Army and Navy Chronicle

Daily National Intelligencer

Globe

Knickerbocker Magazine

Lynchburg Virginian

Military and Naval Magazine

Naval Magazine

New York Times

Niles' Weekly Register

Sailors Magazine and Naval Journal

Southern Literary Messenger

United Service Magazine

Virginia Free Press

Secondary Sources

Adamiak, Stanley J. *The Development of American Naval Logistics.* Lincoln: University of Nebraska Press, 1994.

Agranat, Brina J. "Thorough and Efficient Repair: Rebuilding in the American Sailing Navy." Master's thesis, East Carolina University, 1993.

Albion, Robert. *Makers of Naval Policy, 1798–1947.* Annapolis, MD: Naval Institute Press, 1980.

Allison, William T., Jeffrey Grey, and Janet Valentine. *American Military History: A Survey of Colonial Times to the Present*. 2nd ed. New York: Routledge, 2016.

Allmendinger, David. *Nat Turner and the Rising in Southampton County*. Baltimore: Johns Hopkins University Press, 2014.

An Apology for English Ship-builders: Showing That It Is Not Necessary the Country Should Look to the Navy for Naval Architects. 1833. Accessed May 22, 2020. www.jstor.org/stable/60206267.

Armstrong, Benjamin. *Small Boats and Daring Men: Maritime Raiding, Irregular Warfare, and the Early American Republic*. Norman: University of Oklahoma Press, 2019.

Bauer, K. Jack, ed. *New American State Papers*. Vol. 6, *Administration*. Washington, DC: Scholarly Resources, 1981.

———. "The United States Navy and Texas Independence: A Study in Jacksonian Integrity." *Military Affairs* 34 (April 1970): 44–48.

Bauer, K. Jack, and Stephen S. Roberts. *Register of Ships of the U.S. Navy, 1775–1900*. New York: Greenwood Press, 1991.

Beeler, John. "Maritime Policing and the Pax Britannica: The Royal Navy's Anti-Slavery Patrol in the Caribbean, 1828–1848." *Northern Mariner* 16 (2006): 1–20.

Belohlavek, John M. "Andrew Jackson and the Malaysian Pirates: A Question of Diplomacy and Politics." *Tennessee Historical Quarterly* 36, no. 1 (Spring 1977): 19–29.

———. *Andrew Jackson: Principle and Prejudice*. New York: Routledge, 2016.

———. "Assault on the President: The Jackson-Randolph Affair of 1833." *Presidential Studies Quarterly* vol. 12, no. 3 (Summer 1982): 361–68.

Bennett, Frank. *The Steam Navy of the United States, a History of the Growth of the Steam Vessel*. Pittsburgh: W. T. Nicholson, 1896.

Berube, Claude. "Born to Rule the Seas: The Navy during Andrew Jackson's Presidency and the Genesis of American Naval Power." PhD diss., University of Leeds, 2018.

———. "Budget Battles, Interest Groups, and Relevancy in a New Era: The Ship of the Line USS *Pennsylvania*." Naval Institute *Proceedings* 134 (January 2008): 62–68.

———. "The Crucible of Naval Enlightenment." *Naval History*, October 2014, 58–63.

Berube, Claude, and John Rodgaard. *A Call to the Sea: Captain Charles Stewart of the USS* Constitution. Washington, DC: Potomac Books, 2005.

A Biographical Notice of Com. Jesse D. Elliott. Philadelphia: Russell Jarvis, 1835.

Birkner, Michael. "The Foxardo Affair Revisited: Porter, Pirates, and the Problem of Civilian Authority in the Early Republic." In Watson, *Warfare in the USA*, 165–78.

———. *Samuel Southard: Jeffersonian Whig*. Rutherford, NJ: Farleigh Dickinson University Press, 1984.

Blum, Hester. *The View from the Masthead: Maritime Imagination and Antebellum American Sea Narratives*. Chapel Hill: University of North Carolina Press, 2008.

Bolton, Roy. *Views of Russia and Russian Works on Paper*. London: Sphinx, 2010.

Booraem, Hendrik. *Young Hickory: The Making of Andrew Jackson*. Dallas: Taylor Publishing, 2001.

Borick, Carl P. *A Gallant Defense: The Siege of Charleston, 1780*. Columbia: University of South Carolina Press, 2003.

———. *Relieve Us of This Burthen: American Prisoners of War in the Revolutionary South, 1780–1782*. Columbia: University of South Carolina Press, 2012.

Boyson, V. F. *The Falkland Islands*. Oxford, UK: Clarendon Press, 1924.

Bradford, James. *Command under Sail: Makers of the American Naval Tradition, 1775–1850.* Annapolis, MD: Naval Institute Press, 1985.
———, ed. *Quarterdeck and Bridge: Two Centuries of American Naval Leaders.* Annapolis, MD: Naval Institute Press, 1997.
Brands, H. W. *Andrew Jackson: His Life and Times.* New York: Random House, 2005.
Brockmann, John. *Commodore Robert F. Stockton: Protean Man for a Protean Nation.* Amherst, NY: Cambria Press, 2009.
Brodie, Bernard. *Seapower in the Machine Age.* Princeton, NJ: Princeton University Press, 1941.
Brown, David K. *Before the Ironclad: Warship Design and Development, 1815–1860.* Annapolis, MD: Naval Institute Press, 2015.
Buker, George. "Lieutenant Levin M. Powell, U.S.N., Pioneer of Riverine Warfare." *Florida Historical Quarterly* 47 (January 1969): 253–75.
———. *Swamp Sailors: Riverine Warfare in the Everglades, 1835–1842.* Gainesville: University Press of Florida, 1975.
Burke, Pauline Wilcox. *Emily Donelson of Tennessee.* Knoxville: University of Tennessee Press, 2001.
Burstein, Andrew. *The Passions of Andrew Jackson.* New York: Random House, 2003.
Canney, Donald. *Africa Squadron: The U.S. Navy and the Slave Trade, 1842–61.* Washington, DC: Potomac Books, 2006.
———. *The Old Steam Navy: Frigates, Sloops and Gunboats, 1815–1885.* Annapolis, MD: Naval Institute Press, 1990.
Capowski, Vincent. "The Making of a Jacksonian Democrat: Levi Woodbury, 1789–1851." PhD diss., Fordham University, 1966.
Chapelle, Howard I. *The History of the American Sailing Ships: The Ships and Their Development.* New York: W. W. Norton, 1949.
Cheathem, Mark. *Andrew Jackson: Southerner.* Baton Rouge: Louisiana State University Press, 2013.
———. "The High Minded Honourable Man: Honor, Kindship, and Conflict in the Life of Andrew Jackson Donelson." *Journal of the Early Republic* 27 (Summer 2007): 265–92.
———. *Historical Dictionary of the Jacksonian Era and Manifest Destiny.* Lanham, MD: Rowman and Littlefield, 2017.
———. *Old Hickory's Nephew: The Political and Private Struggles of Andrew Jackson Donelson.* Baton Rouge: Louisiana State University Press, 2007.
———. "The Stubborn Mythology of Andrew Jackson." *Reviews in American History* 47, no. 3 (September 2019): 342–48.
Chisholm, Donald. *Waiting for Dead Men's Shoes: Origins and Development of the U.S. Navy Officer Personnel System, 1793–1941.* Stanford, CA: Stanford University Press, 2001.
Churchman, Joseph, and James Churchman. *Rudiments of National Knowledge: Presented to the Youth of the United States, and to Enquiring Foreigners by a Citizen of Pennsylvania.* Philadelphia: E. L. Cary and A. Hart, 1833.
Cole, Donald B. *A Jackson Man: Amos Kendall and the Rise of American Democracy.* Baton Rouge: Louisiana State University Press, 2004.
Collins, Bruce. *War and Empire: The Expansion of Britain, 1790–1830.* New York: Routledge, 2010.

Cooper, James Fenimore. "Comparative Resources of the American Navy." *Naval Magazine*, January 1836, 21.

———. *The Cruise of the Somers: Illustrative of the Despotism of the Quarter Deck and the Unmanly Conduct of Commander Mackenzie.* New York: J. Winchester, 1844.

———. *Lives of Distinguished Naval Officers.* Philadelphia: E. L. Cary and A. Hart, 1847.

Crawford, Mark, David Stephen Heidler, and Jeanne Heidler. *Encyclopedia of the Mexican-American War.* Santa Barbara: ABC-CLIO, 1999.

Davis, William. *The Greatest Fury: The Battle of New Orleans and the Rebirth of America.* New York: Dutton, 2019.

Dawson, Joseph G. "With Fidelity and Effectiveness: Archibald Henderson's Lasting Legacy to the U.S. Marine Corps." *Journal of Military History* 62.4 (October 1998): 727–53.

Dick, Steven J. "Centralizing Navigational Technology in America: The U.S. Navy's Depot of Charts and Instruments, 1830–1842." *Technology and Culture*, July 1992, 467–509.

Dickinson, H. W. *Educating the Royal Navy: Eighteenth- and Nineteenth-Century Education for Officers.* London: Routledge, 2007.

Drury, Clifford. *The History of the Chaplain Corps, United States Navy.* Vol. 1, *1778–1939.* Washington, DC: Bureau of Naval Personnel, 1948.

Dyson, George. "Charles H. Haswell and the Steam Navy." Naval Institute *Proceedings* 65 (February 1939): 217–30.

Edel, Charles N. *Nation Builder: John Quincy Adams and the Grand Strategy of the Republic.* Cambridge, MA: Harvard University Press, 2014.

Edwards, Judith. *Nat Turner's Slave Rebellion in American History.* New York: Enslow, 2000.

Egerton, Douglas. "Its Origin Is Not a Little Curious: A New Look at the American Colonization Society." *Journal of the Early Republic* 5, no. 4 (Winter 1985): 463–80.

Eland, Ivan. *Recarving Rushmore: Ranking the Presidents on Peace, Prosperity, and Liberty.* Oakland, CA: Independent Institute, 2009.

Feller, Daniel. *The Jacksonian Promise: America, 1815–1840.* Baltimore: Johns Hopkins University Press, 1995.

———. "Politics and Society: Toward a Jacksonian Synthesis." *Journal of the Early Republic* 10, 2 (Summer 1990): 135.

Ferguson, Eugene S. *Truxtun of the Constellation: The Life of Commodore Thomas Truxtun, US Navy, 1755–1822.* Baltimore: Johns Hopkins University Press, 1956.

Flaccus, Elmer. "Commodore David Porter and the Mexican Navy." *Hispanic American Historical Review* 34, no. 3 (August 1954): 365–73.

Gedalecia, David. "Letters from the Middle Kingdom: The Origins of America's China Policy." *Prologue*, Winter 2002. archives.gov.

Germain, Dom Aidan Henry. "Catholic Military and Naval Chaplains, 1776–1917." PhD diss., Catholic University of America, 1929.

Giffin, Donald. "The American Navy at Work on the Brazil Station, 1826–1860." *American Neptune*, 1959, 239–56.

Glenn, Myra. "The Naval Reform Campaign against Flogging, 1830–1850." *American Quarterly*, Autumn 1983, 408–25.

Gough, B. M. "American Sealers, the United States Navy, and the Falklands, 1830–32." *Polar Record* 166 (1992): 219–28

Graham, Gerald S. *The China Station: War and Diplomacy, 1830–1860.* London: Clarendon Press, 1978.

Groen, Mark. "The Whig Party and the Rise of Common Schools, 1837–1854." *American Educational History Journal* 35, iss. 1/2 (Spring/Summer 2008): 251–60.

Gross, Robert, and Mary Kelley, eds. *An Extensive Republic: Print, Culture and Society in the New Nation, 1790–1840.* Chapel Hill: University of North Carolina Press, 2010.

Guttridge, Leonard. *The Commodores.* Annapolis, MD: Naval Institute Press, 1984.

Hacker, Jeffrey, H., ed. *The Early Republic and Rise of National Identity, 1783–1861.* New York: Routledge, 2014.

Hagan, Kenneth. *This People's Navy: The Making of American Sea Power.* New York: Free Press, 1991.

Hamilton, Alexander, James Madison, and John Jay. *The Federalist.* With an introduction by Cass R. Sunstein. Cambridge, MA: Harvard University Press, 2009.

Hargreaves, Mary W. M. *The Presidency of John Quincy Adams.* Lawrence: University Press of Kansas, 1985.

Hart, Joseph C. *Miriam Coffin—or, The Whale Fisherman.* New York: Harper, 1835.

Haywood, Marshall. *John Branch.* Raleigh, NC: Commercial Printing, 1915.

Heflin, Wilson. *Herman Melville's Whaling Years.* Nashville, TN: Vanderbilt University Press, 2004.

Heidler, David S., and Jeanne T. Heidler. *Old Hickory's War: Andrew Jackson and the Quest for Empire.* Mechanicsburg, PA: Stackpole Books, 1996.

Henson, Curtis T., Jr. *Commissioners and Commodores: The East India Squadron and American Diplomacy in China.* Tuscaloosa: University of Alabama Press, 1982.

———. "The United States Navy and China, 1839–1861." PhD diss., Tulane University, 1965.

Hickey, Donald, ed. *The War of 1812: Writings from America's Second War of Independence.* New York: Literary Classics, 2013.

Hidden, Alexander W. *The Ottoman Dynasty: A History of the Sultans of Turkey.* New York: Hidden, 1912.

Hietala, Thomas, R. *Manifest Design: Anxious Aggrandizement in Late Jacksonian America.* Ithaca, NY: Cornell University Press, 1985.

Hoogenboom, Ari, and Herbert Erschowitz. "Levi Woodbury's Intimate Memoranda of the Jackson Administration." *Pennsylvania Magazine of History and Biography,* October 1968, 507–15.

Howe, Daniel Walker. *What Hath God Wrought: The Transformation of America, 1815–1848.* Oxford: Oxford University Press, 2007.

Hudson, Ellis. "Quinine and Atabrine: Development and Present Application." *United States Naval Medical Bulletin* vol. 45, no. 1 (1941): 57–70. Washington, DC: General Printing Office, 1941.

Irving, Washington. *The Complete Works of Washington Irving.* Vol. 1, *1803–1859.* Edited by Wayne R. Kime. Boston: Twayne, 1981.

———. *Complete Works of Washington Irving: Letters 1802–1833.* Edited by Ralph M. Aderman, H. L. Kleinfield, and Jenifer S. Banks. Boston: Twayne, 1978.

Jackson, Carlton. *Bittersweet Journey: Andrew Jackson's 1829 Inaugural Trip.* Morley, MO: Acclaim Press, 2011.

Jampoler, Andrew. *Embassy to the Eastern Courts: America's Secret First Pivot toward Asia, 1832–37*. Annapolis, MD: Naval Institute Press, 2015.

Johnson, Robert Erwin. *Far China Station: The US Navy in Asian Waters, 1800–1898*. Annapolis, MD: Naval Institute Press, 1979.

———. *Thence Round Cape Horn: The Story of the United States Naval Forces on Pacific Station*. Annapolis, MD: Naval Institute Press, 1963.

Kaestle, Carl F. *Pillars of the Republic: Common Schools and American Society, 1780–1860*. New York: Hill and Wang, 1983.

Kaestle, Carl F., Helen Damon-Moore, Lawrence Stedman, and Katherine Tinsley. *Literacy Rates in the United States*. New Haven, CT: Yale University Press, 1991.

Kagan, Robert. *Dangerous Nation*. New York: Alfred A. Knopf, 2006.

Klafter, Craig. "United States Involvement in the Falklands Crisis of 1831–1833." *Journal of the Early Republic* 4, no. 4 (Winter 1984): 395–420.

Kaplan, Fred. *John Quincy Adams: American Visionary*. New York: Harper, 2014.

Katula, Richard. *The Eloquence of Edward Everett, America's Finest Orator*. New York: Peter Lang, 2010.

Kett, Joseph. *The Pursuit of Knowledge under Difficulties: From Self-Improvement to Adult Education in America, 1750–1790*. Stanford, CA: Stanford University Press, 1994.

Kluger, Richard. *Seizing Destiny: How America Grew from Sea to Shining Sea*. New York: Alfred A. Knopf, 2007.

Knox, Dudley. *A History of the United States Navy*. New York: G. P. Putnam's Sons, 1948.

Kotar, S. L., and J. E. Gessler. *The Steamboat Era: A History of Fulton's Folly on American Rivers, 1807–1860*. Jefferson, NC: McFarland, 2009.

Lack, Paul. *The Texas Revolutionary Experience: A Political and Social History 1835–1836*. College Station: Texas A&M University Press, 1992.

Lambert, Andrew. *Battleships in Transition: The Creation of the Steam Battlefleet, 1815–1860*. Annapolis, MD: Naval Institute Press, 1985.

———. *The Last Sailing Battlefleet: Maintaining Naval Mastery, 1815–1830*. London: Conway Maritime Press, 1991.

———. "Preparing for the Long Peace: The Reconstruction of the Royal Navy, 1815–1830." *Mariner's Mirror*, February 1996, 41–54.

———. *War at Sea in the Age of Sail*. London: Cassell Publishing, 2000.

Lambert, Andrew, and Robert Gardiner. *Steam, Steel & Shellfire: The Steam Warship, 1815–1905*. Annapolis, MD: Naval Institute Press, 1992.

Langensiepen, Bernard. *The Ottoman Steam Navy, 1823–1923*. Annapolis, MD: Naval Institute Press, 1995.

Langley, Harold. *The History of Medicine in the Early U.S. Navy*. Baltimore: Johns Hopkins Press, 2000.

———. *Social Reform in the United States Navy, 1708–1862*. Urbana: University of Illinois Press, 1967.

Lanzendörfer, Tim, "Naval Biography, The War of 1812, and the Contestation of American National Identity." In *Warring for America: Cultural Contests in the Era of 1812*, edited by Nicole Eustace and Frederika Teute, 289–91. Chapel Hill: University of North Carolina Press, 2017.

Lardas, Mark. *American Light and Medium Frigates, 1794–1836*. Oxford, UK: Osprey Press, 2008.

Leeman, William P. *The Long Road to Annapolis: The Founding of the Naval Academy and the Emerging American Republic*. Chapel Hill: University of North Carolina Press, 2010.

Leepson, Mark. *What So Proudly We Hail: Francis Scott Key, A Life*. New York: St. Martin's Press, 2014.

Lewis, Charles Lee. *Matthew Fontaine Maury: Pathfinder of the Seas*. Annapolis, MD: Naval Institute Press, 1927.

Libenson, Michael Ben. "American Assistance to the Ottoman Navy in the 1830s and 1840s: A Thesis in History." Master's thesis, Pennsylvania State University, 1987.

Long, David F. *Gold Braid and Foreign Relations: Diplomatic Activities of U.S. Naval Officers, 1798–1883*. Annapolis, MD: Naval Institute Press, 1988.

———. "The Navy under the Board of Naval Commissioners." In *In Peace and War: Interpretations of Naval History*, edited by Kenneth Hagan, 63–78. Westport, CT: Greenwood Press, 1984.

———. *Sailor-Diplomat: A Biography of Commodore James Biddle, 1783–1848*. Boston: Northeastern University Press, 2007.

Longaker, Richard P. "Was Jackson's Kitchen Cabinet a Cabinet." *Mississippi Valley Historical Review* 4, no. 1 (June 1957): 94–108.

Lovette, Leland Pearson. *School of the Sea: The Annapolis Tradition in American Life*. New York: Frederick A. Stokes, 1941.

MacDonald, William. *Jacksonian Democracy, 1829–1837*. Vol. 15 of *The American Nation: A History*. New York: Harper Brothers, 1907.

Madison, Robert, D. "Cooper's Place in American Naval Writing." Presented at the Fourth Cooper Seminar, James Fenimore Cooper: His Country and His Art, State University of New York College at Oneonta, July 1982.

Mahon, John. *History of the Second Seminole War, 1835–1842*. Gainesville: University Press of Florida, 1991.

Marszalek, John, F. *The Petticoat Affair: Manners, Mutiny, and Sex in Andrew Jackson's White House*. Baton Rouge: Louisiana State University Press, 2000.

Martin, H. Lawrence, and Timothy J. Demy. *Sea Padres*. Washington, DC: Office of the Chief of Chaplains, 1984.

McKee, Christopher. *A Gentlemanly and Honorable Profession: The Creation of the U.S. Naval Officer Corps, 1794–1815*. Annapolis, MD: Naval Institute Press, 2001.

———. *Ungentle Goodnights*. Annapolis, MD: Naval Institute Press, 2018.

McLean, David. *Education and Empire: Naval Tradition and England's Elite Schooling*. London: British American Press, 1999.

Meacham, Jon. *American Lion: Andrew Jackson in the White House*. New York: Random House, 2008.

Melville, Herman. "Journal of Melville's Voyage in a Clipper Ship." *New England Quarterly* 2, no. 1 (January 1929): 120–25.

Merrill, James M. "The Asiatic Squadron, 1835–1907." *American Neptune*, April 1969, 106–17.

Mooney, James, ed. *Dictionary of American Naval Fighting Ships*. Vol. 6, *R–S*. Washington, DC: Government Printing Office, 1976.

Morison, Samuel Eliot. *Old Bruin: Commodore Matthew C. Perry*. Boston: Little Brown, 1967.

———. "The Somers Mutiny, 1842–1843." In *Sailor Historian: The Best of Samuel Eliot Morison*, edited by Emily Morison Beck, 181–99. Boston: Houghton Mifflin, 1977.

Morley, Jefferson. *Snow-Storm in August: The Struggle for American Freedom and Washington's Race Riot of 1835*. New York: Anchor, 2012.

Murray, Williamson. *The Shaping of Grand Strategy: Policy, Diplomacy and War*. Cambridge: Cambridge University Press, 2011.

Nagel, Paul. *John Quincy Adams: A Public Life, a Private Life*. New York: Alfred P. Knopf, 1997.

Napier, Charles. *An Account of the War in Portugal between Don Pedro and Don Miguel*. Cambridge: Cambridge University Press, 2013.

Novotny, Patrick J. *The Press in American Politics, 1787–2012*. Santa Barbara, CA: Praeger, 2014.

Nye, Russell. "Freedom Road: Nat Turner." *Centennial Review of Arts and Science* vol. 1, no. 3 (Summer 1957): 242–63.

Padfield, Peter. *Maritime Power and the Struggle for Freedom: Naval Campaigns, 1788–1851*. New York: Overlook Press, 2006.

Paine, Lincoln P. *Warships of the World to 1900*. Boston: Mariner Books, 2000.

Pearce, George F. *The U.S. Navy in Pensacola: From Sailing Ships to Naval Aviation, 1835–1930*. Gainesville: University Press of Florida, 1980.

Penmer, Gayle. "Historic Notes and Documents: The Mix Diaries." *Florida Historical Society* 82, no. 2 (Fall 2003): 191–218.

Philbrick, Nathaniel. *Sea of Glory*. New York: Penguin, 2004.

Philbrick, Thomas. "Cooper and the Literary Discovery of the Sea." *Canadian Review of American Studies* 20 (Winter 1998): 35–45.

———. *James Fenimore Cooper and the Development of American Sea Fiction*. Cambridge, MA: Harvard University Press, 1950.

Pletcher, David M. *The Diplomacy of Annexation: Texas, Oregon, and the Mexican War*. Columbia: University of Missouri Press, 1973.

Poe, Edgar Allan. "Critical Notices: Maury's Navigation." *Southern Literary Messenger* 2.7 (1836): 454–55.

Pratt, Julius. "The Origin of Manifest Destiny." *American Historical Review* 4 (1927): 795–98.

Rathbun, Richard. "The Columbian Institute for the Promotion of Arts and Sciences." Smithsonian Institution bulletin. Washington, DC: Government Printing Office, 1912.

Ray, Angela G. *The Lyceum and Public Culture in the Nineteenth Century United States*. East Lansing: Michigan State University Press, 2005.

Remini, Robert. *Andrew Jackson*. New York: Harper, 1999.

———. *Andrew Jackson and the Course of American Democracy: 1833–1845*. Baltimore: Johns Hopkins University Press, 1984.

———. *Andrew Jackson and the Course of the American Empire: 1767–1821*. New York: Harper and Row, 1977.

———. *Andrew Jackson and the Course of American Freedom: 1822–1832*. Baltimore: Johns Hopkins University Press, 1998.

———. *Andrew Jackson and His Indian Wars*. New York: Viking, 2001.

———. *The House: The History of the House of Representatives*. New York: Collins, 2007.

———. *The Life of Andrew Jackson*. New York: Harper Perennial Press, 2010.

Rentfrow, James. *Home Squadron: The U.S. Navy on the North Atlantic Station*. Annapolis, MD: Naval Institute Press, 2014.

Richardson, James. *A Compilation of the Messages and Papers of the Presidents*. Vol. 2, *1817–1833*. Authority of Congress, 1900.

———. *A Compilation of the Messages and Papers of the Presidents*. Vol. 3, *1833–1841*. Authority of Congress, 1900.

Rippy, J. Fred. *Joel R. Poinsett: Versatile American*. Durham NC: Duke University Press, 1935.

Robichaud, Andrew A. *Animal City: The Domestication of America*. Cambridge, MA: Harvard University Press, 2019.

Rodger, N. A. M. "Navies and the Enlightenment." In *Essays in Naval History*, Surrey, UK: Ashgate, 2009.

Rohr, John. "Constitutional Foundations of the United States Navy." *Naval War College Review* 45 (Winter 1992): 68–84.

Rorabaugh, W. J. *The Alcoholic Republic: An American Tradition*. New York: Oxford University Press, 1979.

Satz, Ronald N. *American Indian Policy in the Jackson Era*. Norman: University of Oklahoma Press, 1975.

Schama, Simon. *The Embarrassment of Riches: An Interpretation of Dutch Culture in the Golden Age*. New York: Alfred A. Knopf, 1988.

Schlesinger, Arthur M., Jr. *The Age of Jackson*. Boston: Little, Brown, 1945.

Schroeder, John H. *Commodore John Rodgers: Paragon of the Early American Navy*. Gainesville: University Press of Florida, 2006.

———. "Jacksonian Naval Policy 1829–37." In *New Aspects of Naval History: Selected Papers from the 5th Naval History Symposium*, edited by Department of History, US Naval Academy, 121–27. Baltimore: Nautical and Aviation Publishing Company of America, 1985.

———. *Matthew Calbraith Perry: Antebellum Sailor and Diplomat*. Annapolis, MD: Naval Institute Press, 2001.

Schueller, Malina Johar. *U.S. Orientalisms: Race, Nation, Gender and Literature, 1780–1890*. Ann Arbor: University of Michigan Press, 1998.

Semmel, Bernard. *Liberalism and Naval Strategy: Ideology, Interest, and Sea Power during the Pax Britannica*. Boston: Allen and Unwin, 1986.

Shoemaker, Nancy. *Native American Whalemen and the World: Indigenous Encounters and the Contingency of Race*. Chapel Hill: University of North Carolina Press, 2015.

Shoemaker, Raymond L. *Diplomacy from the Quarterdeck: The United States Navy in the Caribbean, 1815–1830*. PhD diss., Indiana University, 1976.

Shulman, Mark Russell. *The Emergence of American Sea Power: Politics and the Creation of US Naval Strategy*. Berkeley: University of California Press, 1990.

Skaggs, David Curtiss. *Thomas MacDonough: Master of Command in the Early U.S. Navy*. Annapolis, MD: Naval Institute Press, 2002.

Smith, Gene. *Thomas ap Catesby Jones: Commodore of Manifest Destiny*. Annapolis, MD: Naval Institute Press, 2000.

Smith, Mark A. *Engineering Security: The Corps of Engineers and Third System Defense Policy, 1815–1861*. Tuscaloosa: University of Alabama Press, 2009.

———. "Engineering Slavery: The U.S. Army Corps of Engineers and Slavery at Key West." *Florida Historical Quarterly* vol. 86, no. 4 (Spring 2008): 505–10.

Sobocinski, Andre. "A Navy Surgeon on Olympus." *Navy Medicine* vol. 96, no. 5 (2005): 28–31. Bethesda, MD: Navy Medical Department, 2005.

Somit, Albert. "Andrew Jackson as Administrator." *Public Administration Review* vol. 8, no. 3 (Summer 1948): 188–96.

Sondhaus, Lawrence. *Naval Warfare, 1815–1914*. Abingdon, UK: Routledge, 2001.

———. *Navies of Europe: 1815–2002*. London: Longman, 2002.

Sprout, Harold, and Margaret Sprout. *The Rise of American Naval Power, 1776–1918*. Annapolis, MD: Naval Institute Press, 1990.

Stapleton, Augustus Granville. *George Canning and His Times*. London: John W. Parker and Son, 1859.

Starr, Paul. *The Creation of the Media: Political Origins of Modern Communications*. New York: Basic Books, 2004.

Stephens, John G. *Incidents of Travel in Greece, Turkey, Russia and Poland*. Dublin: Curray, 1839.

Sterns, Worthy Putnam. "The Foreign Trade of the United States from 1820 to 1840." *Journal of Political Economy* vol. 8, no. 4 (September 1900): 452–90.

Stevens, Kenneth. *Border Diplomacy: The Caroline and McLeod Affairs in Anglo-American-Canadian Relations, 1837–1842*. Tuscaloosa: University of Alabama Press, 1989.

Stockton, Charles. "The Declaration of Paris." *American Journal of International Law*, 3 (1920): 356–68.

Symonds, Craig. *Confederate Admiral: The Life and Wars of Franklin Buchanan*. Annapolis, MD: Naval Institute Press, 1999.

———. *Historical Atlas of the U.S. Navy*. Annapolis, MD: Naval Institute Press, 1995.

———. *Navalists and Antinavalists: The Naval Policy Debate in the U.S., 1785–1827*. Wilmington: University of Delaware Press, 1980.

Taylor, Andrew, and Aine Kelly, eds. *The Idea of America*. New York: Routledge, 2013.

Temperley, Harold. *The Foreign Policy of Canning, 1822–27*. Abington, UK: Frank Cass, 1925.

Thiessen, William. *Industrializing American Shipbuilding: The Transformation of Ship Design and Construction, 1820–1920*. Gainesville: University Press of Florida, 2006.

Tocqueville, Alexis de. *Democracy in America*. New York: Harper and Row, 1965.

Toll, Ian. *Six Frigates: The Epic History of the U.S. Navy*. New York: W. W. Norton, 2006.

Tomblin, Barbara Brooks. *Bluejackets and Contrabands: African Americans and the Union Navy*. Lexington: University of Kentucky Press, 2009.

Tomlins, Christopher. "In Nat Turner's Shadow: Reflections on the Norfolk Dry Dock Affair of 1830–1831." *Labor History* 33 (1992): 494–518.

Tucker, Spencer. *Andrew Foote: Civil War Admiral on Western Waters*. Annapolis, MD: Naval Institute Press, 2002.

———, ed. *Enduring Controversies in Military History*. Santa Barbara, CA: ABC-CLIO, 2017.

Vale, Bernard. *A War betwixt Englishmen: Brazil against Argentina, 1825–1830*. New York: I. B. Tauris, 2000.

Valle, James E. *Rocks and Shoals: Order and Discipline in the Old Navy, 1800–1861*. Annapolis, MD: Naval Institute Press, 1996.

Vandervort, Bruce. *Indian Wars of Mexico, Canada, and the United States, 1812–1900*. New York: Routledge, 2006.

Vargas, Mark A. "The Military Justice System and the Use of Illegal Punishments as Causes of Desertion in the U.S. Army, 1821–1835." *Journal of Military History* vol. 55, no. 1 (January 1991): 1–19.

Warshauer, Matthew. *Andrew Jackson and the Politics of Martial Law: Nationalism, Civil Liberties, and Partisanship*. Knoxville: University of Tennessee Press, 2006.
Watson, Harry L. *Liberty and Power: The Politics of Jacksonian America*. New York: Hill and Wang, 2006.
Watson, Richard L., Jr. "Congressional Attitudes toward Military Preparedness, 1829–1835." *Mississippi Valley Historical Review* 34, 4 (March 1948): 611–36.
Watson, Robert. *Affairs of State: The Untold History of Love, Sex, and Scandal, 1789–1900*. New York: Routledge, 2012.
Watson, Samuel J. *Jackson's Sword: The Army Officer Corps on the American Frontier, 1810–1821*. Lawrence: University Press of Kansas, 2012.
———. *Peacekeepers and Conquerors: The Army Officer on the American Frontier, 1821–1846*. Lawrence: University Press of Kansas, 2013.
———, ed. *Warfare in the USA, 1784–1861*. Abingdon, UK: Routledge, 2017.
Weddle, Kevin. *Lincoln's Tragic Admiral: The Life of Samuel Francis Du Pont*. Charlottesville: University of Virginia Press, 2005.
Wettemann, Robert P., Jr. *Privilege vs. Equality: Civil-Military Relations in the Jacksonian Era, 1815–1845*. Santa Barbara, CA: ABC-CLIO, 2009.
Williams, David. *A People's History of the Civil War*. New York: New Press, 2005.
Williams, S. T. *The Life of Washington Irving*. Oxford: Oxford University Press, 1935.
Wootton, David, ed. *The Essential Federalist and Anti-Federalist Papers*. Indianapolis: Hackett, 2003.

Index

Page numbers in italics refer to figures.